TECHNOLOGY AND INFORMATION TRANSFER

A Survey of Practice in Industrial Organizations

RICHARD S. ROSENBLOOM
Professor of Business Administration
Harvard University

FRANCIS W. WOLEK
Assistant Professor of Industry
University of Pennsylvania

Formerly Research Associate in Business Administration
Harvard University

DIVISION OF RESEARCH
GRADUATE SCHOOL OF BUSINESS ADMINISTRATION
HARVARD UNIVERSITY
BOSTON · 1970

Library of Congress Catalog Card No. 70–119550
ISBN 0–87584–087–6

PRINTED IN THE UNITED STATES OF AMERICA

Table of Contents

List of Tables

List of Figures

Foreword

IN 1963 RESEARCH AT THE HARVARD BUSINESS SCHOOL was initiated by Richard Rosenbloom and Frank Wolek to describe the process by which technical information is communicated and used. National and corporate attention had been increasingly drawn to the interaction between scientific and technological advance and social change. This problem was considered to be of considerable practical importance at that time, as it still is. This volume is a product of that effort. The research focuses upon the flow of technical information across organizational lines in the research and development operations of large industrial corporations. The basis of the work is a body of survey data collected from 2,000 engineers and scientists in 13 establishments of 4 corporations, and from 1,200 members of the Institute of Electrical and Electronics Engineers. The data describe instances in which respondents acquired useful technical information from sources outside their immediate circle of colleagues. The analysis is descriptive in character, following a functional approach in which the use of various means of information transfer is considered in relation to the purposes of technical work.

The authors undertook this project to increase our understanding of technical communications in research and development and to enrich comprehension of the complex interactions between the technical, organizational, and cost factors involved. They discuss the effects which management may have, within an organization, on the process of information transfer, and the need for managers and students of the process to take into account the interplay of personal values, task requisites, and the structure of formal and informal social groups.

This research project has depended on many sources for its financial support. In its early stages, the Committee on Space of the American Academy of Arts and Sciences encouraged Professor Rosenbloom by lending its support. A grant to the School from the National Science Foundation (Grant GN-305) provided additional funds for the authors to carry on two more years of research. Funds given by The Associates of the Harvard Business School completed the project's financial requirements. On behalf of the School, I should like to express gratitude and thanks to these sources for their valued support of this research project.

LAWRENCE E. FOURAKER
Dean

Soldiers Field
Boston, Massachusetts
June 1970

Preface

RESEARCH AND DEVELOPMENT is part of the booming "knowledge industry" that is becoming a major force in American enterprise. It is an industry that feeds on knowledge in order to create more of it. A vast diversity of disciplines, purposes, and institutional settings is encompassed by the simple rubric "R&D," but there is a unifying characteristic in the work of the many scientists and engineers so variously engaged. Their main task is problem solving: systematically, creatively, building on the work of peers and predecessors. Most of this problem-solving work is performed in industry and directed toward the practical objectives of a firm or public agency. How well those objectives are met, and at what cost, depends to an important degree on the ability of the working scientist and engineer to acquire the technical information needed to do his job. This book describes an aspect of the process by which that is accomplished.

The great outpouring of potentially relevant technical information and the increasingly interdisciplinary character of work complicate the task of acquiring needed technical information. The work is done within organizations: the free flow of information may be impeded by organizational and institutional boundaries, i.e., among divisions of large corporations, between the corporations, and between government agencies, private industry, and the universities. By attempting a systematic description of the flow of information in industrial R&D, we hope to contribute to a better understanding of the nature of the process of technical communication, of the factors influencing it, and, ultimately, of the means by which it can be strengthened. Our findings should be of interest not only to those responsible for the design of technical information systems but, more important, to those managers responsible for the work of scientists and engineers in R&D.

This research originated as part of a series of studies of the problem of technology transfer begun in 1963 for the Committee on Space of the American Academy of Arts and Sciences (under a NASA grant) along with concurrent studies of R&D management at the Harvard Business School. A user survey, analyzing the flow of information in a single large corporation, was conducted at the start of 1964 as part of the American Academy's project. This first survey served to develop effective research instruments, methods of analysis, and some working hypotheses. A generous grant to the Harvard Business School from the National Science Foundation, Office of Science Information Service, permitted us to expand and intensify the survey effort and to broaden the research to examine industrial R&D activity through related case studies. This book reports the findings of the survey research.

We received encouragement, insightful comment, and helpful cooperation in good measure from many individuals and organizations. The companies which cooperated in this study and the over 2,000 engineers and scientists who served as respondents must remain anonymous. Obviously, this study would have been impossible without the time and thoughtful attention so generously given by the respondents, officers, staff, and supervisors in those companies. In addition, we owe a debt of gratitude to the officials of the Institute of Electrical and Electronics Engineers (IEEE) and the respondents to our mail surveys for their help in making the IEEE survey possible.

Most of the statistical analyses for this study were performed at the Harvard University Computing Center. We made heavy use of the contingency table program written by David Armor and the Multiple Statistical Analyzer developed by Kenneth J. Jones. To the staff of the Center and the several programmers and consultants who assisted, we also owe a great deal. Additional computational work was performed at the University of Pennsylvania's Computer Center with programs written by Terry Overton. James Miller III, who was then at MIT's Project MAC, wrote an original set of programs for and contributed significantly to the regression analysis presented in Chapter 3.

Many colleagues provided valuable and welcome assistance to help resolve troublesome aspects of the design of the study or to help clarify and improve various reports of its findings. At the Harvard Business School, Professors Paul Lawrence, Howard Raiffa, and Lewis Ward provided notable assistance. Throughout the period of work on this project, our efforts were strengthened by continuing interaction with our associates William Abernathy, Leonard Harlan, Curtis McLaughlin, and William Yost. Encouragement and help from outside Harvard was provided in special measure by Thomas Allen, Donald Marquis, and Richard Orr. The help of two colleagues was of special importance to us. Raymond Bauer maintained a continuing interest in the project from its inception, for which he was largely responsible, and made helpful comments on an earlier draft of this manuscript. Bertrand Fox, who served as Director of the Division of Research for most of the active period of this project, provided useful guidance, unfailing encouragement, and unbelievable patience in the face of the profound slowness of our progress toward completion.

Dean Lawrence E. Fouraker, serving as Director of Research during the final portion of our efforts, provided, as in other matters, a rare blend of stimulation, support, and wise counsel. We wish also to thank former Dean George P. Baker and Senior Associate Dean George F. F. Lombard for their help, and in particular for arranging our schedules to provide the time needed to conduct this study.

Authors require the assistance of many skilled persons in the preparation of any published work. We are indebted, no less than others, to a large number of such persons, beyond our ability to identify them all individually. To begin with, the staff of the Harvard University Printing Office provided efficient professional assistance in preparing the questionnaires used in our surveys. Three secretaries contributed substantially to the project during its duration; we offer special thanks to Phyllis Furst, Marilyn Moses, and Paula Sabloff. Much of the typing was done by Theresa C. Sparks and by the staff directed by Rose Giacobbe. Barbara Mello provided considerable editorial assistance at an important juncture. We are grateful, finally, to Hilma Holton, who competently oversaw the conversion of our manuscript into this book in a way that was faithful to our interests yet undemanding of our attention.

Our debt to so many others does not, of course, discharge our responsibility for the findings and conclusions of this research, which we hereby cheerfully and explicitly acknowledge.

RICHARD S. ROSENBLOOM
FRANCIS W. WOLEK

Boston and Philadelphia
June 1970

Technology and Information Transfer

CHAPTER 1

Introduction

TECHNOLOGY PROVIDES THE MEANS by which man interacts with his environment in the satisfaction of his needs. The essence of technology is cognitive, not material. Whether simple mechanism or complex system, whether "hard" tool or "soft" program, at its core technology is the embodiment of man's understanding of natural processes. Each progressive advance in technology expresses some man's ability to grasp an essential principle and society's ability to preserve and transmit that insight through time and across space. The effort described here is a study of one aspect of the latter ability; it is a study of the transfer of information in the development of new technology.

BACKGROUND

Continuing change in technology is a central fact of modern times. What is unique about our time, however, is not the fact of change but its scale and character. The important innovations in technology are now "science-based," that is, they spring from recent extensions of scientific knowledge, and their development requires a scale and complexity of effort of a new order. Examples can readily be seen in new systems for transportation, communications, and energy production, and in spectacular developments in military and space technology.

The modern symbol of the mutually reinforcing interactions among science, technology, and industry is the industrial research and development laboratory. At present, two-thirds of all research and development work in the United States is performed by private industry. Although federal support has been prominent in the growth of R&D expenditures, fully half of the work in industry is privately funded. As is well known, the growth of organized science and technology in our generation has been remarkable. In 1941 there were only 90,000 research scientists and engineers employed in all sectors, including industry, government, and universities. R&D spending that year was less than $1 billion. By 1965 industry alone employed some 360,000 R&D scientists and engineers in work costing $15 billion annually.[1]

By the very nature of science and technology, information transfer must represent an important function in the conduct of technical work.[2] The cumulative body of technical knowledge is so great, and its rate of growth so rapid, that no individual can be acquainted with more than a small portion of the whole. In any specialized field, whether scientific discipline or technological art, significant advances can arise from widely scattered individual efforts or the work of several organizations. No one can avoid a dependence on predecessor and contemporary efforts elsewhere. There is interdependence across fields; the essential unity of scientific and technological understandings of nature means that almost any bit of knowledge may prove to be relevant to almost any particular need. These characteristics of knowledge and practice in

---

[1] National Science Foundation, *Basic Research, Applied Research, and Development in Industry, 1965* (1967).

[2] One can cite rough quantitative measures of the importance of this activity. Studies of the behavior of scientists and engineers have shown that a large fraction of their time is devoted to the acquisition or dissemination of information, that is, to listening and talking, reading and writing. For the nation as a whole some $500 million is spent each year for research, development, and innovation in information systems and sciences and for the operation of technical information systems.

technical fields create the need for an effective coupling across the necessary bounds of specialized disciplines and organizations. The information transfer system, the subject of this study, serves that function.

Aims and Methods of the Study

This study has both an immediate and a broader purpose. It is, in the first instance, a study of the means by which scientists and engineers in industrial research and development organizations acquire technical information useful in their work yet unavailable from their own knowledge or that of their immediate colleagues. The matter of most immediate interest, then, is the question of how scientists and engineers become aware of and build upon the experience of fellow professionals employed in other fields of work and in other organizational settings. As an empirical study, it must have an extremely narrow focus; our data pertain only to a particular activity, information transfer, in a particular work setting, the industrial R&D organization. Yet there is a broader purpose behind our interest in that activity and that setting. Both the specific activity and the institutions in which we study it are important components of the process by which new technology is created. The last is the object of our broader interest; we hope to contribute to a better understanding of the nature of the work from which technological progress ensues.

There are two questions around which the study is organized: (1) "By what means does information flow between technical groups?" and (2) "Under what circumstances does information transfer take place by one means and under what circumstances by another?" In an empirical study each of the general concepts in these questions must be given operational meaning. Categories must be defined for "means," measures developed for "circumstances," and so forth. The reader will find that the choices we made in framing these categories and the questions explored in the analysis of the data were shaped substantially by the character of our broader aims.

In brief, we sought to provide an empirical basis for understanding the process of information transfer in relation to the nature of organized R&D work. As the research questions imply, we accomplish this through an analysis of the relationships between the means by which R&D professionals acquire information and the circumstances of their work. In reviewing the existing discourse on this subject, we found that principal concern centered on two points: (1) the extent to which information transfer permits new work to build on experience built up outside the immediate work locale and (2) the specific kinds of documentary and interpersonal sources which effect the transfer of such information. These two provide the basis for key distinctions made in the analysis of our data. Organizational factors also are prominent in our approach to the study of this process. We have studied work which is performed in mission-oriented organizations, almost all of them of large size, and we have defined information transfer in terms of the most basic unit of organization, the group of immediate colleagues reporting to the same supervisor.

The technique used to collect data descriptive of the flow of information was based on the sampling of incidents of actual information exchange in the organizations we studied. The unit of data analysis is a description provided by each respondent of an instance in which he acquired information useful to his work from a source outside his immediate circle of colleagues. Each respondent was asked to describe his most recent instance, spelling out the circumstances leading up to the acquisition of the information and the source from which he obtained the substance of the information. The data were obtained by surveys in which we reached nearly 1,900 engineers and scientists in 13 establishments of 4 large corporations, and 1,200 members of the Institute of Electrical and Electronics Engineers.

There are three principal goals in our analysis of the survey data:

(1) To identify the relative incidence of the use of specified means of acquiring information.
(2) To analyze these findings in relation to varying circumstances of work.
(3) To infer from that analysis certain fundamental characteristics of the technical work itself.

In the analysis, the characteristics of the source of information represent the main dependent variables. We consider them in relation to aspects of the social and technological context — specifically, the personal background of the respondent, the setting within which he works, and characteristics of the task to which the information is applied.

The Prevailing Discourse

Certain issues set the framework within which we will examine the process of information flow. Although this is a descriptive and not a prescriptive study, it aims to present that description in terms relevant to the manager for the resolution of policy issues in his own organization.

There is a substantial literature concerned with policies bearing on the transfer of technical information, and it reveals significant differences in views. For, although there is widespread acceptance of the importance of facilitating free access to the best of knowledge from whatever source, that sort of agreement does not imply a common evaluation of strategies and methods for prompting such access. While information may always be a "good thing," its acquisition is not without cost. For this reason, there is room for significant disagreement about the appropriateness of particular institutional policies relating to mechanisms for the transfer of information and particular patterns of individual behavior relating to its acquisition.

The desirability of an open system of information flow sustaining the essential unity of both science and technology is, of course, widely accepted by technical professionals. The common view is clearly expressed in the 1963 report, *Science, Government and Information,* by the President's Science Advisory Committee (PSAC):[3]

Science and technology can flourish only if each scientist interacts with his colleagues and his predecessors, and only if every branch of

---

[3] President's Science Advisory Committee, *Science, Government, and Information* (1963), p. 7.

science interacts with other branches of science. In this sense science must remain unified if it is to remain effective.

What sort of system for information transfer can create this kind of unity? Given the general belief that information useful to a given task may arise from work in various disciplines and across a wide geographic area, there is an understandable emphasis on the need for each worker to seek information from the best sources and to keep "current" on a special area of expertise. It follows from this that any specialized group can expect to depend on the work of other groups to meet important information needs in its own work. Thus, one element of the professional expectation about information services and practices is that they will facilitate the individual's ability to acquire information from sources beyond his own knowledge and that of his immediate colleagues.

This prevailing professional view implies certain expectations about appropriate media for information transfer. The PSAC report, for example, goes on to state that "the ideas and data that are the substance of science and technology are embodied in the literature." [4] Information published in the archival journals of the established disciplines is, by its nature, universally accessible; it also possesses a status in the technical community overshadowing ephemeral publications or informal exchanges between individuals. Dr. James Shannon, Director of the National Institutes of Health, said (in 1963) that "information is not viewed [in the biomedical research community] as definitive or examined seriously until it has survived the editorial processing procedures and found its level of excellence in the hierarchy of journals." [5] Yet other media necessarily figure in the flow of information to professionals. Dr. Shannon, in the same statement, stressed the value of scientific meetings and argued that the prepublication delays of journals are offset by the "extremely important role" of interpersonal exchange.

Not surprisingly, then, questions relating to documentation have

---

[4] *Ibid.*

[5] Federal Council for Science and Technology, *Proceedings:* Second Symposium on Technical Information and the Federal Laboratory, 1964, pp. 12–16.

traditionally dominated policy-oriented discussions of the role of information in technological progress. A comprehensive government-sponsored study of "National Systems for Scientific and Technical Information" by the System Development Corporation (SDC) was devoted almost exclusively to document processing systems.[6] SDC's report offers the brief observation that "oral and other informal forms of communication play a vital but vaguely understood role in the work of scientists, practitioners, and engineers."

As the last statement tacitly acknowledges, professionals both in science and in engineering do rely quite heavily under certain circumstances on local and interpersonal sources of information; yet the nature of the reliance remains, as the SDC comment puts it, "vaguely understood." In part, this probably reflects a feeling in many quarters that this sort of communication is not a good thing. It seems more reasonable to us, however, to assume that prevailing practices represent more than a simple matter of irrational concern for convenience or of ignorance of alternatives. Some sensible and well-informed people believe that they represent effective patterns of behavior. For example, an engineering manager who is highly regarded by his peers and subordinates told us, "When I see one of my men reading a professional journal I know he is wasting his time." A noted nuclear physicist stated in a published interview:[7]

> Theoretical physics is a subject in which . . . speech is the important thing, not the written word. You have to go around and talk with people and be in contact if only to learn that this particular mess of papers here on my desk is rubbish and these others are the important thing. . . .

Thus while there is widespread acceptance of the importance of the free flow of information across organizational and other "artificial" boundaries, there is no comparable agreement on the means by which that flow should be sustained. A given situation may be evaluated in contradictory ways. For example, if one finds

---

[6] Committee on Scientific and Technical Information, *Recommendations for National Document Handling Systems in Science and Technology*, 1965.

[7] Abdus Salaam in an interview published in *International Science and Technology*, December 1964, p. 70.

that engineers in a particular department rarely use a technical library close at hand, the finding constitutes a problem only if one believes that the library "ought" to be used. If this is considered to be a problem, one might ask either "What's wrong with the library?" or "What's wrong with the engineers?" As Christopher Scott observed in the conclusion of a study which found that literature was used very little by certain engineers, "*a priori* this might reflect a fault in the literature, a fault in the technologist, a fault in the facilities for making the literature available to the technologist, or there may be no fault at all — it might be a good thing." [8]

Scott goes on to conclude "if there is a fault, most of it lies in the technologist." But contrasting interpretations are also common. The PSAC report says, for example, "A simple but urgent suggestion to authors is to refrain from unnecessary publication. The literature has been and always will be cluttered with poor and redundant articles." [9] As another example, a study of library use in a given setting might disclose ways in which the library was "at fault" and suggest methods of improving its services.

To return to Scott's original assessment, "there may be no fault at all — it may be a good thing." [10] It is not necessary to infer

---

[8] C. Scott, with L. T. Wilkins, *The Use of Technical Literature by Industrial Technologists*, 1960, pp. 64–65.

[9] President's Science Advisory Committee, *Science, Government, and Information* (1963), p. 25.

[10] In fact, it is difficult to find evidence that the information transfer system, as a whole, is performing inadequately. Clearly, there are significant problems in elements of the system — inadequacies in the literature, shortsighted policies of employers, and so forth — and new ideas and new technology create important opportunities for improvement. Yet the people who do the work find that their needs are being met. In the DOD User Needs Study (by Auerbach Corporation, Phase 1, Final Technical Report, 2 volumes, Philadelphia, Pa., May 1965 — AD 615 501 and 615 502), for example, the majority of respondents did not consider that they had a problem finding information when they needed it (Volume 2, pp. 6–43). Other evidence in the same study shows why users feel content with their present practices: in almost every case the first source consulted provided at least some of the needed information (Volume 2, pp. 6–38), in more than 95% of the cases the user obtained all the information he needed within the necessary time span (Volume 2, pp. 6–105).

that what is observably true is a problem merely because it does not mesh with some loosely stated expectations. Neither is it necessary to assume that what is, is good because it is. Instead, one should ask under what circumstances certain patterns of information transfer are found, inquire why this might be so, and use this explanation to gain a better understanding of the nature of R&D work. This is our approach, and it is readily identified with a "functional" point of view.[11]

Our aim is to understand the nature of the information transfer system by analyzing the use of different sources as functional elements of a single system. We are willing to assume, in the absence of contrary evidence, that the behavior of our subjects is effective, in the sense of fulfilling their own objectives, and that an observed pattern of information flow is the result of sensible collective efforts to satisfy a complex set of goals within resource constraints.

THE RESEARCH CONTEXT

During the past 20 years, as the scale and importance of science and technology have grown so markedly, the study of technical communication has expanded apace. Having described, in general terms, the aims and content of our study and the main issues shaping our approach to the study, we turn now to a consideration of this body of related research, in order to set ours in its proper context. Related work is that which is concerned with the "empirical study of scientific communications in process among scientists

---

[11] It may be that more mature appraisal will show that the process is at fault, but such a conclusion would be premature in the absence of an understanding of why competent men behave as they do. An excellent discussion of the utility of this approach to research in communications is contained in R. A. Bauer, "The Obstinate Audience," 1964. Bauer has argued elsewhere ("Problem Solving in Organizations: A Functional Point of View," 1963) ". . . that the task of the social sciences is to discover the rationale of behavior whether or not it appears to be rational . . . man's behavior can seldom be as wise as we generally assume it can; and it is wiser than we think it is."

and technologists in the course of their professional activities." [12]

The work of others is significant for its methodological contributions, as well as substantive findings. Specifically, the means employed by other researchers made an important contribution to the design of the data-gathering methods used in this study. For many years, however, the development of this field was hampered by the difficulty of developing reliable data-collection techniques. Thus, a study of previous work also warned us what not to do.[13]

A landmark in the development of better methods of research was the publication of a *Review of Studies in the Flow of Information Among Scientists*, written by Herbert Menzel at the Bureau of Applied Social Research at Columbia University. Although Menzel's more recent review points out that studies continue to be performed in ignorance of the lessons in this field,[14] the last decade has brought a substantial number of soundly based and widely relevant studies.

Turning to substantive characteristics, we find that studies in this field can be grouped broadly into two categories, according to the type of system whose performance is of principal concern to the investigator. One type of study is concerned primarily with the performance of an information service or system of services which provides information to particular populations of scientists or engineers. The main goal in such studies is an evaluation of the effectiveness (and often also the cost) of available information

---

[12] The defining phrase is taken from H. Menzel, "Scientific Communication: Five Sociological Themes," 1966. Menzel's review and that by W. J. Paisley, "The Flow of (Behavioral) Science Information — A Review of the Research Literature," 1965, provide comprehensive interpretive studies of work in this field.

[13] For example, many early studies were based on diary records or on questionnaires using rather abstract questions. Diaries have since been shown to be unreliable, primarily because subjects are either unwilling or unable to interrupt the normal flow of work and employ them reliably and validly. Questionnaires asking "Do you generally use —— ?" or "What percent of your time do you spend doing —— ?" have also turned out to be misleading because they ask the subject to form a general observation from heterogeneous experiences over a vaguely defined time span and permit both witting and unwitting bias to enter the answers.

[14] H. Menzel, "Information Needs and Uses in Science and Technology," 1966.

media and services in meeting specified user needs. The alternative research orientation concentrates on the impact of information on the technical job — as opposed to the impact of the information service on the user — and examines the work of science and technology itself, as it is affected by the information available to professional practitioners. Research of the latter type often is primarily concerned with the social system within which information is generated and used. Our research fits within this tradition, being concerned with an exploration of the interface between the task and the information, rather than that between the user and the information service. It will be useful, however, to consider briefly the characteristics and contributions of both approaches.

Studies of the needs of scientists and engineers and the effectiveness of actual and proposed information services in meeting those needs constitute the larger proportion of research in this field. These are commonly referred to as "User Studies." Their main orientation is clinical in the sense that they tend to be concerned with the study of symptoms which indicate the health of a given system of services providing information. Such studies often are closely tied to efforts to revamp existing information storage and retrieval systems or to design new ones. An experienced practitioner in this field has said that "defining the compromise that will satisfy the greatest segment or discommode the smallest segment of the user group" is the principal objective behind studies "to determine and define information use patterns and requirements." [15]

User studies have a definite value, as well as some important limitations. The clinical approach is often appropriate for the evaluation and development of a specific service in a particular setting. Although many methodological difficulties have handicapped studies of this sort in the past, most have been overcome in recent work. In studies designed and interpreted by professionals, management may find extremely useful ideas about possibilities for the improvement of existing services.[16]

---

[15] S. Herner, "The Determination of User Needs for the Design of Information Systems" in *Information Systems Workshop*, Conference sponsored by the American Documentation Institute, Washington, D.C., May 29–June 1, 1962 (Spartan Books, 1962), p. 47.

[16] Not all professionals in the design of information systems have acknowl-

Yet clinical studies of information needs and uses are inevitably bound by present and practical information services and particular sets of users. As Herbert Menzel has said:[17]

> Information needs are not synonymous with either the demands or the conscious wants of information users. It is not the information that users are aware of wanting that counts, nor even the information that would be good for them, but rather the information that would be good for science — for the progress of scientific research.

A concern for the impact of information on the primary task of the user, i.e., on the progress of science and technology, represents an important change of stance in the study of "technical communication in process." The focus of study shifts away from the systems for storage and distribution and toward the systems for consumption of technical information. The implication of that shift is that the researcher should consider the value of information services in terms of their net contribution to the work of science and technology, rather than merely their effectiveness or cost in meeting specific needs for information.

Some very imaginative work has been directed at the development of techniques which would account for the value of information in terms of its contribution to work in science and technology. Two main approaches to the direct, quantitative measurement of value can be identified. One assumes that the benefit to technical work is maximized when the time and total cost of acquiring information are minimized.[18] The other confronts the direct product of the technical work (e.g., scientific papers, designs,

---

edged the value of studies of user needs. For example, M. Taube ("An Evaluation of Use Studies of Scientific Information," December 1958) states: "(My) inevitable conclusion is that Use studies have no value as direct guides to the design of information systems any more than consumer acceptance or rejection is a guide to the value of the Salk vaccine."

[17] "Can Science Information Needs Be Ascertained Empirically?" New York University, June 29, 1966.

[18] As in the studies of time allocation by M. H. Halbert and R. L. Ackoff, "An Operations Research Study of the Dissemination of Scientific Information," 1959.

program definitions) and obtains informed professional judgments of merit.[19]

In other studies the value of information is approached indirectly. That is, without employing quantitative measures of the costs of particular efforts or the worth of particular outcomes, the research is directed at an analysis of the functions of information in the work of science and technology.[20] These studies therefore are concerned with the *ways* in which information is of value in technical work, rather than with direct measures of the *magnitude* of such value. As is implied by our earlier discussion, that is the approach taken in this study. We seek to define the ways in which alternative means of obtaining information are functional to the purposes of the work of research and development.

In the foregoing discussion we have tried to make clear where our work stands in the developing stream of research on the use of technical information. We should also make clear the several specific characteristics of our work which differentiate it from other empirical studies of information flow. These include the nature of the population studied, the size and character of the sample employed, the method of data analysis, and the interpretive framework for analysis. No prior study represented a comparable range of situations within industrial settings. Few employed sample sizes large enough to lend conviction to generalizations beyond the population immediately under study.[21] Furthermore, aided by the size and composition of the sample, we use multivariate techniques for data analysis which permit us to consider simultaneously a number of aspects of the user's situation. By so doing we can trace in our analysis the implications of our initial conception of information transfer as a process of interrelated components.

Since there are other studies of the sort we conducted, it is ap-

[19] See M. W. Martin, "The Measurement of Value of Scientific Information," 1963, and Thomas J. Allen, "Performance of Information Channels in the Transfer of Technology," 1966, pp. 87–98.

[20] H. Menzel, "The Information Needs of Current Scientific Research," 1964.

[21] The Department of Defense User Needs Studies, conducted first in government laboratories and then among industrial contractors, represent a comparable body of research. These studies began at about the same time as ours, and their data are broadly consistent with ours.

propriate to ask what they tell us that is relevant to our own purposes and also how the findings of this study might differ from
the others. At the level of relatively simple descriptive results, those
who have investigated information transfer among R&D engineers
and scientists have consistently found the same general pattern:
Most engineers and scientists in mission-oriented organizations use
informal media and local sources to meet most of their needs for
technical information.

In the Department of Defense User Needs Study, for example,
a survey of engineers and scientists in the laboratories of industrial
defense contractors showed that in five out of seven cases in
which it was necessary to search for information, the man first
consulted a source within the company or his own or departmental
files.[22] "Project Hindsight," a Defense Department study of the
origins of information and ideas that were important in the development of 20 successful weapons systems, showed that in 70%
of the cases personal contact was the medium by which the information was introduced into the using organization.[23]

The general finding that local and informal sources serve most
information needs in mission-oriented R&D implies the corollary
conclusion that infrequent use is made of the formal literature and
the information services that exist to provide access to that literature. There is, of course, considerable evidence that technical
libraries, information centers, and related services are used relatively infrequently. In a British study, for example, a total of 600
scientists who used one of 25 selected technical libraries on a
test day were asked about details of literature which, had they
discovered it earlier, would have caused them to alter their research in some way. In 33% of the cases cited, the source of the
discovery was a colleague's recommendation; 1% had their at-

---

[22] North American Aviation, Inc., "DOD User Needs Study, Phase II,"
November 1966.

[23] As additional examples, see C. Scott, "The Use of Technical Literature
by Industrial Technologists," 1962; S. Herner, "Information Gathering Habits
of Workers in Pure and Applied Science," January 1954; M. H. Halbert and
R. L. Ackoff, "An Operations Research Study of the Dissemination of Scientific Information," 1959; Organization for Economic Cooperation and Development, *Technical Information and the Smaller Firm*, 1960; and T. J. Allen,
"Sources of Ideas and Their Effectiveness in Parallel R&D Projects," 1965.

tention drawn by library staff. The same men, asked to rank 12 ways of getting information in order of usefulness, gave highest ranking to following citations, keeping up with current literature, using journal indexes, and conversation with colleagues. Using the library card index and asking the librarian ranked 11th and 12th out of 12. These respondents were all library users in the first place.[24] As another example, the Director of Research of the Plastics Department of the Du Pont Company has described the operation of a mechanized coordinate indexing system which covered 9,300 documents (more than half were patents; the remainder comprised research reports and market analyses). The system had cost $218,000 to develop and was costing $60,000 annually to operate. After four years of operation, one-third of the client group of 300 to 400 men had never used the system; among users, 2% of the people accounted for 17% of the questions asked, while the average was 1½ questions per user.[25]

As we have indicated above, at the simple descriptive level our findings are consistent with those of other studies of information flow. While the larger sample size and greater range of settings may permit somewhat finer detail in the description and perhaps somewhat greater confidence in generalization, none of the main results diverges significantly from what might have been expected. Similarly, those experienced in the work of science and engineering will find that these results largely match their own expectations based on generalizations from direct experience. It should not be surprising that a systematic investigation of what goes on in information transfer should produce results consistent with everyday observation; after all, the data we collected were derived from the same events that the manager sees every day.

Yet despite general agreement on what goes on in information transfer, there is a definite lack of consensus on what it means. And this is why we choose to pursue our analysis beyond the level of simple description of the sources and needs for technical infor-

[24] C. W. Hanson, "Research on Users' Needs: Where Is It Getting Us?" 1964.

[25] Federal Council for Science and Technology, *Proceedings:* Second Symposium on Technical Information and the Federal Laboratory, 1964, pages 41–45.

mation. By a functional analysis of the system for information transfer — specifically, by an analysis of the *means* used to acquire information in relation to the *purposes* of the work in which it is employed — we seek to develop a deeper insight into the nature of R&D work, as well as into the workings of the information transfer process.

CHAPTER 2

Survey Methods

THIS IS A DESCRIPTIVE STUDY OF BEHAVIOR. The data we collected are concerned only with what people actually *do*, rather than with what they ought to do or would like to do, or with what they think or feel about what they do. The activity in question transpires when a person acquires technical information that is useful to his work and is obtained from a source other than his own knowledge, or that of his immediate circle of colleagues. The people we studied are scientists or engineers doing research and development work in industrial organizations. Data were collected by means of a self-administered questionnaire that elicited descriptions of a single incident of information transfer and answers to other questions pertaining to the work setting, the task to which the information was applied, and the personal characteristics of the respondent.

To interpret our findings, the reader should be aware of the characteristics of the methods of data collection. In this chapter we discuss the scope of the survey, key variables in the data, the assumptions underlying our analysis, and the validity of our findings.

As indicated previously, survey questionnaires were collected in 1965 from 1,900 engineers and scientists in 13 establishments of four large corporations and from 1,200 members of the Institute of Electrical and Electronics Engineers (IEEE).[1] In each of the participating corporations we selected several establishments with significant R&D operations and through the senior technical manager distributed questionnaires to all "professionals" employed in R&D work either as individual contributors or as firstline supervisors.[2] The IEEE survey was conducted by mail using a sample drawn from the membership of four specialized professional groups within the organization.

Each subject was asked:

> Please think of the most recent instance in which an item of information, which you received from a source other than someone in your immediate circle of colleagues, proved to be useful in your work. Answer the following questions with reference to that instance.

A page of instructions preceding this request spelled out the interpretation to be put on each of the key phrases. A typical questionnaire, including instructions, is presented in Appendix A.[3]

The questionnaire is in two main parts; the first, as the appendix shows, pertains to characteristics of the instance itself. The first four questions and the seventh deal with the circumstances leading up to acquisition of the information — specifically, the respondent's prior awareness of his need for the information; whether he had been seeking it and, if so, why; and his use of a lead in

---

[1] We conducted an exploratory survey in 1964 in a large diversified corporation, using an earlier version of our questionnaire. More than 1,200 instances were described by 430 respondents in five locations, including the central R&D laboratory. For a description of the survey and a discussion of these generalizations, see C. P. McLaughlin, R. S. Rosenbloom, and F. W. Wolek, "Technology Transfer and the Flow of Technical Information in a Large Industrial Corporation," 1965.

[2] Those with professional degrees but employed in other ways (e.g., as salesmen) were excluded from the survey, while those without professional degrees but doing jobs ordinarily considered professional were included. Computer programmers were included in the sample in those organizations working in the data processing field.

[3] The questionnaires varied in detail from place to place. Some of these differences are noted at the beginning of Appendix A. Of significance here is the fact that in each organization the phrase "immediate group of colleagues" was specifically defined by reference to the organization's own term for the smallest organized work group (e.g., section, department, unit).

identifying a source. In Questions 5 and 6 the respondent is asked to identify the specific source and the medium of communication involved. Two questions inquire about the respondent's experience with the relevant technical field. Finally, Questions 10–13 concern the function of the information, identifying the nature of the task to which it was applied and the character of its effects on that task. A second section of the questionnaire, headed "Background Questions," elicits information about the respondent himself and, in the mail questionnaires, about his work setting.

KEY VARIABLES

The central variables in our analysis describe the manner in which the respondent acquired the substance of the information concerned.[4] In classifying these responses we distinguish first between the use of interpersonal and written media, and then between types of sources as follows:

If the information was obtained by interpersonal communication, we distinguish among sources in terms of the place of employment of the informant. Engineers or scientists employed in the same engineering department or research laboratory as the respondent are called "local" sources; other employees of the same corporation are called "other corporate" sources; finally, all persons employed outside the respondent's corporation are termed "external-interpersonal" sources.

The three categories involving written media are differentiated on another basis. All written sources that originated in the respondent's corporation are grouped as "company documents." Published trade magazines or documents originating in other organizations, i.e., catalogs, technical reports, etc., are grouped together as "trade" documents. Finally, all published books or published technical or scientific journals are grouped as "professional" documents.

---

[4] Our data define only the proximate sources of the information. For example, if A learned something from B and B had learned it originally from a journal, A's report of the incident should specify B as the source and interpersonal communication as the medium. Person B, describing the acquisition of the same substantive information, would have identified a professional journal as the source.

Identifying the immediate source of the information and the medium that ties the user to that source tells only part of the story of information transfer. We should also consider the specific circumstances leading up to the acquisition of information from that source. The principal distinctions made here concern the user's intent in seeking the information and his prior awareness of the need for it. Three main categories serve to classify the various circumstances reported. Most often, of course, the respondent had been searching specifically for the information he actually acquired. There are a number of cases, however, in which the initiative came from someone other than the user. A colleague or other person may have pointed out some useful information, either spontaneously or in response to an earlier indication of interest by the respondent. Finally, there is that important class of events in which a scientist or engineer acts with the intention of developing his general competence, rather than of finding the particular problem-solving information that he in fact obtained. In our analysis these three classes of circumstances are termed "specific search," "pointed out," and "general competence," respectively.

The second set of principal variables pertains to the circumstances of work. These measures deal with the technological content of the work (identified by Question 10), the organizational setting (primarily, whether in a central research laboratory or a department of an operating division), the professional background of the individual (in science or in engineering), and certain other personal characteristics of the user (including seniority, professional activity, reading habits, and so forth).

SCOPE OF THE STUDY

A definition of the scope of the study is implicit in the structure of the survey data, as just defined. Here we make the limits of the data explicit, in a brief discussion of scope.

This is a study of the *transfer* of *technical* information, and does not address other aspects of the acquisition of information in technical organizations. Thus we are not concerned with communication of other sorts of information, such as knowledge of budgets, schedules, or personnel, although they are obviously also of great significance to the effective functioning of a technical organization.

Nor do we deal with all of the means for acquiring technical information. Obviously, much of the information for R&D work comes directly from the professional's own knowledge or out of the results of observation and experimentation in the lab. These sources of information are not treated here. As noted in the introduction, we have defined the transfer of technical information in a very specific way. It is the process by which substantive information in science or technology, originating in one organizational setting, is acquired and put to use by engineers and scientists doing research and development within another organizational unit. Our data exclude instances limited to the exchange of information between immediate colleagues. Although these exchanges are important within the total information system and are interrelated with the information transfer process as we have defined it, they raise questions other than those studied here, and their study would require research methods quite different from those used here.[5] The questionnaire design also excluded certain immediate sources of technical data habitually employed in technical work, principally handbooks. This, too, was done largely on methodological grounds,[6] although there is empirical evidence that the exclusion is of minor import.[7]

---

[5] The unstructured nature of the exchange of information between men working in close proximity on similar or related tasks poses a different order of questions for the design of research and the collection of data. The pertinence of such research to questions of work group organization and supervision is clear. We suggest, however, that the main questions which form the background of this study (discussed in Chapter 1) can be addressed adequately by working solely with the intergroup transfer of information.

[6] To many engineers (and scientists) the use of such sources of standard data is so much a part of their routine that they fail to distinguish or even recognize specific incidents. The survey method we used, when developed for the behavioral sciences, was described as the "critical incident" method, since it was meant to focus on clearly distinct events in people's lives. Accordingly, we have found that the method is best used to study incidents involving information which plays a direct part in decision-making and problem-solving processes. The ability of the technique to catch incidents in which people use a source of information for routine verification or for a detailed "memory" of things they already know is much more subject to doubt.

[7] All sources of technical information (including "own knowledge") were

Assumptions in Analysis of the Data

The fundamental assumption underlying our use of the "incident" as the unit of analysis is that the "most recent instances" described by a group of respondents represent an unbiased sample from the set of all instances in which those respondents acquired such information over a substantial time period. Hence, for any group of related respondents we take the relative incidence of a characteristic of the reported instances as an estimate of the incidence of that characteristic in all cases in which those respondents acquired such information.

For example, 28% of the respondent chemists in a pharmaceutical laboratory described instances in which they had obtained information from a source within their own firm, 55% reported books or professional journals as the source, and 17% reported using other sources outside the company. To translate this statement about a particular set of responses into generalizations describing characteristics of the transfer of technical information, over time, to chemists in that organization, four assumptions must be made. They are (a) that our respondents' descriptions of sources (in their answers to Questions 5 and 6) were valid, (b) that they selected "the most recent instances" without bias, (c) that their pattern of use of various sources to obtain information does not fluctuate significantly over moderate periods of time, and (d) that the nonrespondent chemists in the lab were not significantly different from the respondents.

Of course, as we have already stated, our primary interest lies neither with simple statements about the relative incidence of various sources in instances of information transfer, nor with statements limited to the specific organizations included in the survey.

---

included in the DOD User Needs Studies. In the study among industrial contractors, handbooks and manuals (the latter surely more prevalent in military work) were cited in approximately 5% of the cases. The report implies that most were not of the published variety, i.e., they fell in the category of informal documents reported in our data. Thomas J. Allen (*Proceedings*, 20th National Conference on the Administration of Research, 1966, p. 121), commenting on his own studies of information use, said, "We found little use of handbooks."

Our aim is to formulate, in operational terms, a set of widely relevant predictive propositions defining the relationships between aspects of the technical task, the individual's orientation toward work, and the nature of the organizational contexts, on the one hand, and consistent patterns of behavior in acquiring technical information, on the other.

In order to support these more general propositions, of course, we must also be able to make specific inferences of the type exemplified by the example of the pharmaceutical lab given above. In other words, we must first be able to describe the transfer of information in a given setting. By observing differences in characteristics of information transfer as we find them in different groups, we may be able to identify systematic associations between those differences and some characteristics of the individuals, their tasks, or the work setting. For example, we will find that the incidence of use of sources outside the respondent's own firm is greater among respondents who are authors of professional articles or books than among those who are not, greater among those who are scientists than among those who are engineers, and greater among those who are doing research than among those who are engaged in design and development. Finally, we wish to draw a third type of inference, in which observations of the second type can be linked together into even more general propositions based on underlying characteristics of the task, the individual, or the situation.

Can our data validly support this type of analysis? Table 2-1 suggests a hierarchy of levels of data aggregation for various levels of generality in inference. Our example from the pharmaceutical lab was at level 3. Five requisite assumptions are noted in relation to that and lower levels.[8]

We have tested assumptions (a) through (e), identified in Table 2-1, in a number of ways and are satisfied that they are valid for the data we are using and for the inferences we will draw. The validity of questionnaire answers and the selection of incidents were examined through a program of interviews before and

---

[8] These include the four assumptions (a) through (d) noted above, plus a fifth which acknowledges the possibility of sampling error within a unit. The last arises primarily in relation to the IEEE Groups, since there was 100% coverage and a high response rate in most corporate establishments.

TABLE 2-1

Levels of Analysis

	From these data: *Answers to Questions 1–13 describing an "instance," as reported by:*	*We make inferences about:* *Characteristics of the acquisition of technical information by:*	*Based on these key assumptions:*
Level 1	A respondent.	The respondent in that one event.	(a) Validity of categories used in the questionnaire.
Level 2	A closely related group of respondents, e.g., immediate colleagues.	That group of men, over time.	(b) Unbiased selection of "most recent instance" by respondents. (c) Stability of process characteristics over moderate periods of time.
Level 3	Respondents in one establishment or IEEE group.	All professionals in that establishment or group.	(d) Nonrespondents not significantly different from respondents. (e) No significant sampling error.
Level 4	All respondents.	Scientists and engineers doing R&D in large industrial organizations under company funding.	(f) Other organizations not significantly different from those in survey.

after the survey. The response rate in the corporate surveys was very high: 80% overall and more than 90% in six establishments. Mail and telephone follow-ups on IEEE nonrespondents and the use of two mailings for the survey made it possible to test non-respondent bias, time stability, and sampling error. The details of these tests are set forth in Appendix B.

As a consequence of these tests, we have discarded certain data and interpret others only in a limited way.[9] With these amendments, however, the survey data may be considered sound for inferences up to the third level of generality (i.e., organizations surveyed). Extending them to the highest level implies a further assumption, that of comparability between organizations, on which the reader may make his own judgment. Although the relative utilization of various sources may vary significantly in some other settings, it seems likely to us that the direction and importance of the effects of the factors being studied would not be substantially different.

Sources of the Data

For the reader to be able to determine the relevance of our data in the specific contexts with which he is concerned, he must know more about the sources of our data. Specifically, he must have some knowledge of the factors that will determine whether another organization is judged to be significantly different from those we studied.

Surveys in Corporations

No single organization can be taken to "typify" the setting of industrial R&D. In each of the firms in which we conducted surveys, we met with senior corporate executives to select a small but varied group of establishments which would represent important categories of R&D work in the company. These organizations are described in Table 2-2.

---

[9] Responses obtained from computer programmers generally are excluded from our analysis because of special difficulties that appeared in those cases. Data from Question 7 (lead) and Questions 11–13 (function) were discarded because interviews showed that their meaning was not consistently interpreted.

TABLE 2-2

Participating Organizations

Name and Description	Size and Response	
	Question-naires Dis-tributed *	*Percent Usable Returns*
MEDCO The central research laboratory in a company that manufactures pharmaceuticals. It is located at the principal site of manufacturing operations.	269	91%
BASIC CORPORATION Two laboratories at a single site performing the central R&D functions for divisions of a corporation producing basic materials. Manufacturing operations are widely dispersed throughout the country.	252	72%
DELTA CORPORATION A very large, diversified corporation, in which seven establishments in five operating units were selected for study. All seven establishments were located at the same site as their associated manufacturing operations. The corporation also operates central research and engineering laboratories. The data are grouped as follows in our analysis:		
POLY Three small chemical-development labs associated with operating units at separate locations within a 100-mile radius.	138	91%
MECHO The design and engineering departments of a unit manufacturing a single type of large mechanical equipment, primarily for a single industry.	203	83%
CONTROLS The engineering organizations for two separate operations manufacturing electronic controls and control systems.	226	97%

Table 2-2 (continued)

Name and Description	Size and Response	
	Questionnaires Distributed *	Percent Usable Returns
EDP The engineering department concerned with product and systems development for an organization producing electronic data-processing equipment.	347	66%
DATA CORPORATION A very large corporation whose principal business is the design, manufacture, and sale of data-processing equipment and systems. Four laboratories within a 200-mile radius were selected for study.		
RESEARCH The corporation's fundamental research laboratory.	455	70%
ADVANCED DEVELOPMENT Responsible for the development of advanced concepts and systems.	180	89%
EQUIPMENT An organization concerned with design and development of EDP hardware; located at an important manufacturing site for the company.	360 **	75%

* Equal to the number of professionals employed, except as noted.
** A stratified sample drawn from more than 1,000 professionals was employed.

Since we hoped to analyze the effects of variations in personal and situational factors, we tried to achieve enough variety on important dimensions to assure statistically meaningful returns. At the same time, we selected only those establishments within which the technical work was directed primarily at contributions to the creation of new technology. None of the establishments or groups filled a strictly service-oriented role; all were actively and directly engaged in research or innovative development and design. Our survey sample, furthermore, does not include professionals employed in technical service work such as spectroscopy, technical writing, and quality testing, nor does it include so-called "production" engineers or designers who concern themselves primarily with refinement of existing products, trouble-shooting services, or routine adaptation of products to particular customer needs.

Few if any of the respondents worked in isolation from sources of relevant information. All establishments were large enough so that an individual would have access to a number of colleagues and experts performing work ancillary to his own responsibilities. None of the firms or establishments, furthermore, pursued policies that appeared likely to have significantly restricted information flow. All but two of the establishments had a technical library close at hand; some of the libraries were sizable (see Table 2-3).

It should be noted, however, that two types of R&D organizations — one common, the other less so — are not represented at all in our data. One is the government contracting laboratory; the other is the small technically oriented firm.

The most common setting for R&D in industry is, of course, the organization working under contract to the federal government. As we noted earlier, federal support, primarily from the military and space agencies, accounts for one-half of all the R&D work performed by industry. We chose not to investigate this area because of the special requirements of those R&D tasks and the security arrangements restricting information flow.[10] Thus, most of the establishments in our survey worked only on company-sponsored projects, and none had any significant amount of contract work.

We also excluded R&D in small firms, largely because of the statistical nature of our research technique. This is a less important exclusion. More than 70% of the engineers and scientists engaged in company-funded R&D work are employed by fewer than 100

---

[10] The Department of Defense, furthermore, has recently supported larger-scale studies of information flow in government laboratories and industrial contractors (Auerbach Corporation, "DOD User Needs Study, Phase I," May 1965; North American Aviation, Inc., "DOD User Needs Study, Phase II," November 1966).

TABLE 2-3

Availability of Technical Library

Business	Library Staff		Library Holdings*			
	Professional Staff	Other Staff	Books	Bound Periodicals	Journals	Technical Reports
MEDCO	(Library in same building, no data published)					
BASIC A**	1	1	1,200	800	150	500
B	3	10	25,000	1,000	475	10,000
DELTA CORPORATION						
POLY A†	2	2	1,500	200	125	n.r.
B	1	1	—	2,400	120	8,000
C	no library					
MECHO††	5	7	n.r.	n.r.	600	n.r.
CONTROL A		1	—	270	28	
B	no library					
EDP DATA CORPORATION	1	2	2,700	350	400	2,000
Research	5	7	18,000	8,000	800	n.r.
Advanced Development	3	2	4,000	400	300	5,000
Equipment	1	2	8,000	1,170	230	n.r.

* n.r. indicates data not reported.

** The letters, A,B,C, designate separate establishments, grouped together in our analysis.

† No library as part of R&D organization; this library was in the same establishment, but was administratively separate.

†† In addition to the main library serving the entire Division, the engineering department maintained a large technical data center with a staff of four.

companies, each of which employs 500 or more scientists and engineers.[11]

[11] National Science Foundation, *Basic Research, Applied Research, and Development in Industry, 1965* (1967), Table 43.

We believe that both the participating organizations and the respondent individuals fairly represent the work of proprietary R&D in the laboratories of large industrial firms. Our four firms are among the very largest in the country. The establishments include, principally, operations in the machinery, chemical and allied products, and electrical equipment and communications industries. Those three industries accounted for nearly one-half of all company funds for R&D in 1965, and in those three, the large firms (those with total employment of 5,000 or more) provided nearly 80% of the aggregate R&D funds in each industry.[12] As to personal characteristics, our sample includes a greater proportion of scientists than does the national population of R&D scientists and engineers, and the distribution of educational attainments seems slightly higher than the norm. Neither difference is large. In other respects, as a group the respondents are probably typical of the total population. Statistics on the personal characteristics of respondents are tabulated in Appendix C.

IEEE Surveys

Our approach to data collection was the same in the IEEE surveys as in the surveys conducted within corporations. The questionnaires used in all the surveys were similar in their essential features (see Appendix A). The method of sample selection, however, was entirely different in the case of the IEEE, resulting in a set of respondents who were substantially different in a number of respects from those participating in the surveys in corporations. For the IEEE surveys, furthermore, questionnaires were distributed through the United States mails, rather than through channels within the subject's own organization.

The IEEE is an international professional society created by the merger of the American Institute of Electrical Engineers and the Institute of Radio Engineers. Our survey sample was not selected from among the membership at large; it was restricted instead to IEEE members who had affiliated themselves with at least one of four selected professional groups within the organization. The structure of the IEEE includes a number of these specialized

---

[12] *Ibid.*, Table 13.

groups. Each brings together those members who share a common interest in a technical field defined in terms of some distinct body of theory or specific knowledge within the broader domain of Electrical Engineering. Each group publishes quarterly *Transactions*, holds symposia and meetings, sponsors local chapters in various areas of the country, and assesses membership dues in addition to those charged to IEEE members at large. In other words, each group performs many of the functions of a professional society for those who choose to maintain awareness of and contribute to developments in the special area of knowledge which defines its purpose. Affiliation with a group is voluntary; approximately two-thirds of the general membership is not affiliated with any group; those who join a group tend to belong to more than one.

The sample to which we mailed questionnaires was drawn from the membership of four groups: (1) Industrial Electronics and Control Instrumentation (IECI); (2) Circuit Theory; (3) Power; and (4) Electronic Computers.[13] These four were selected because they represented fields in which a significant number of respondents in the corporate surveys were also engaged. They were of sufficient size, individually and in aggregate, to yield an adequate number of responses.

The data from the IEEE surveys can be taken as representative of the population of group members engaged directly in engineering work within organizations.[14] Group members, however, constitute a special population, differing in significant respects from the broader population of Electrical Engineers engaged in industrial R&D. The set of Electrical Engineers in our corporate surveys cannot be assumed to be representative of the latter population, either. Respondents in the corporate surveys were selected

---

[13] The mechanism for this was simple. The contents of every eighth tray in the file of address plates for all IEEE members in the United States were sorted according to group membership. A questionnaire was mailed to everyone belonging to at least one of the selected groups.

[14] The survey design and the reliability and validity of the results are discussed in Appendix B. Returns from IEEE members were screened to exclude those who were self-employed or not primarily engaged in technical work. See the cover letter for the first mailing and the screening questions on page two of the second questionnaire, in Appendix A.

by the arbitrary selection of certain organizations; they do, however, constitute a broad sampling of engineers, including a number of IEEE Group members.

The Framework of Analysis

In the introduction we stated the purpose of our data analysis to be the identification of relationships between the sources used in the transfer of information and the circumstances of the work in which the information is used. We have defined the key variables in our description of information transfer and of the circumstances of work. The major premise of our analysis is that there should be an empirically testable association between one and the other.

Correlation and causation are two different things, of course. Variables that are found to be associated cannot always be presumed to be linked by a direct causal chain. For example, an observed association between authorship of published papers and greater use of sources outside the firm will not be interpreted to suggest that the condition of authorship is a cause of the observed pattern (although it may be in some particular cases). Instead, authorship of professional publications is seen as one of several indicators of a personal orientation toward the development of professional competence. That orientation, representing a general posture toward work, or any one of the variables we use as indicators of that orientation, may play a causal role in a particular situation.

Even allowing for the qualification above, the question of inferring causality from the data remains difficult. For example, reciprocal relationships undoubtedly exist between characteristics of individuals and of situations. One aspect of a man's orientation toward his work is his personal commitment to developing substantive skills. This commitment is expressed in part by his having pursued his education toward higher degrees, but the orientation itself is also developed and shaped by education, since advanced study tends to strengthen this kind of personal orientation. A single descriptor, furthermore, may be tied to more than one intervening variable. The data to be discussed in Chapter 3, for example, show significant differences between scientists and engineers. But we cannot tell from this simple comparison whether the differences in use of information sources are a consequence of differences between science and engineering as categories of work, between scientists and engineers as professional populations, or between scientific and engineering organizations. Given these conceptual difficulties and ambiguities, the goal of seeking cause and effect relationships would present difficulties that we do not seek to overcome and limitations that we do not wish to accept.

Hence, the goal of this study is not to establish theoretical and universal generalizations. We seek only to present a predictive framework using concepts which are reasonably meaningful to management on the one hand and related to theory in the behavioral sciences on the other. In this effort we have been guided by an orientation well expressed by Bruner, Goodnow, and Austin[15] as follows:

> Science and common-sense inquiry alike do not discover the ways in which events are grouped in the world; they invent ways of grouping. The test of an invention is the productive benefits that result from the use of invented categories. . . .

In summary, what we seek is predictive guidelines which have some probabilistic validity. Individual managers, in their attempts to use these guidelines, will need to gather that information from their staffs and information specialists; this would allow them to determine the particular circumstances present in their organization and thus the applicability of the guidelines we offer.

---

[15] J. S. Bruner, J. J. Goodnow, and G. A. Austin, *Study of Thinking*, 1962, page 7.

CHAPTER 3

Sources for Information Transfer

IN THE INTERPRETATION OF THE SURVEY DATA, the first question to be posed is the simplest one: to what extent do the respondents utilize alternative sources for information transfer? Our premise, of course, was that that question should not be posed generally, but rather in relation to particular circumstances of work. The latter are considered within three categories, namely, the organizational context, the characteristics of the task, and the personal characteristics of the user. Before discussing these, however, it will be useful to review the data in the light of two other salient factors: the user's profession and the particular circumstances leading up to acquisition of the information.

SCIENTISTS AND ENGINEERS

There are consistent differences in the utilization of alternative sources of information between scientists and engineers employed in the organizations we studied.[1] Scientists tend to make substantially more use than do engineers of sources outside the

---

[1] The classifications "scientist" and "engineer" were made on the basis of answers to the question on professional field which appears as Question 20 in Part II of the corporate survey questionnaire (see Appendix A).

corporation, a difference which is especially marked in respect to the use of professional journals and books. Among scientists, sources within their own corporation provide information in only one-third of the instances, as opposed to the strong majority — typically three-fifths — of such instances reported by engineers. The data concerning sources of information are summarized in Figure 3.1 for all scientists and engineers responding to the survey.

FIGURE 3.1

AGGREGATE SURVEY RESULTS[a]

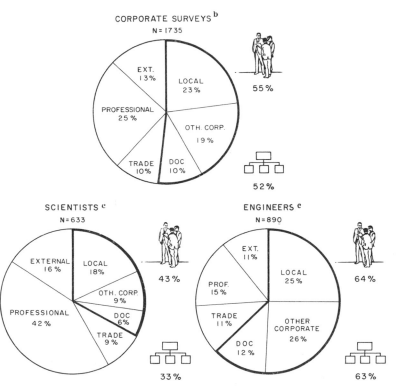

[a] See Notes for Figure 3.1, page 36.

[b] This group includes computer programmers employed in the EDP Division of DELTA and in three divisions of DATA Corporation.

[c] Based on respondents' answers to Question 20 (Appendix A, Part II).

The format of these charts will be followed in most of the figures in this chapter and the next. The categories for sources follow the definitions given in Chapter 2, as follows:

I. Sources within the respondents' own company (enclosed by heavy line):
    (A) *Interpersonal* communication:
        (1) *Local* source — an engineer or scientist employed in the same establishment.
        (2) *Other corporate* source — another person employed by the same corporation.
    (B) *Written* Media:
        (3) *Documents* — any written source originating in the same corporation.

II. Sources outside of the company:
    (A) *Written* Media:
        (4) *Trade* documents — suppliers' catalogues, trade magazines, unpublished technical reports, etc.
        (5) *Professional* documents — published books, journal articles, or conference papers.
    (B) *Interpersonal* communication:
        (6) *External*-interpersonal — communication with any person employed outside the firm.

The percentage of responses in each category is given within the pie chart. The subtotals for interpersonal media (1, 2, and 6), and corporate sources (1, 2, and 3), are given at the right of the pie chart under the upper and lower sketches, respectively.

The figure also shows, for completeness, the aggregate survey results — including scientists, engineers, and some programmers. The reader will recognize that the aggregate outcome is strongly affected by the balance between the various professional groups in our sample.

Although these differences between professional groups could be the result of an unusual pattern of use on the part of only one class of scientists or engineers who happened to be given great weight in our sample, that situation is not apparent when we split these broad fields into more specific categories, as in Figure 3.2.[2]

---

[2] A word on the basis of this disaggregation is in order. We do not, of course, have a probability sample of members of any professional field. Electrical and mechanical engineering and chemistry are well-recognized fields for

Among these respondents, differences among fields within science or engineering appear small in relation to the differences between scientists and engineers in aggregate. The principal exception to this generalization is that chemical and metallurgical engineers report a greater incidence of use of professional documents and, correspondingly, a lesser reliance on sources within their own firms, than do mechanical and electrical engineers. The latter two, in turn, are quite similar in the pattern of sources reported.

PRIOR CIRCUMSTANCES

The respondents' descriptions of the circumstances[3] leading up to the acquisition of information provide some indication of the heterogeneity of events in the information-transfer process. In aggregate, slightly fewer than half the instances resulted from specific search by the respondent. In nearly one-third of the cases, the information was acquired because someone pointed it out. Finally, in about one-fifth of the cases, the respondent's intention in seeking information had been to develop his general competence rather than to acquire some particular knowledge. With instances of specific search in the minority, it is clear that information transfer is not just a problem of information retrieval. In the transfer of technical information in industrial laboratories, information looking for the man seems to be nearly as frequent an occurrence as the man seeking information.

In fact, in many cases the engineer or scientist will not be aware

---

which we had data from several sizable groups in several organizations. The other categories are arbitrary. The remaining engineers, two-thirds of whom were in chemical and metallurgical fields, were grouped together. Most of the remaining scientists were accounted for by two categories, into which we place them for convenience of presentation: the "life scientists," all of whom were employed in the MEDCO laboratory, and physicists and mathematicians, employed in EDP or the DATA Corporation (termed "computer scientists").

[3] As defined in Chapter 2, we grouped these into three categories, according to whether the information had been sought for the specific use to which it was put, had been pointed out by someone else, or had been acquired in the course of "competence development" activities such as keeping up with or reviewing a technical field.

FIGURE 3.2

Information Sources by Field

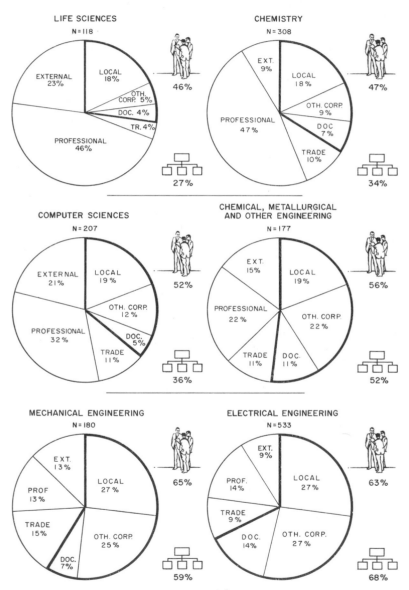

LIFE SCIENCES
N=118

EXTERNAL 23%
LOCAL 18%
OTH. CORP. 5%
DOC. 4%
TR. 4%
PROFESSIONAL 46%
46%
27%

CHEMISTRY
N=308

EXT. 9%
LOCAL 18%
OTH. CORP. 9%
DOC. 7%
TRADE 10%
PROFESSIONAL 47%
47%
34%

COMPUTER SCIENCES
N=207

EXTERNAL 21%
LOCAL 19%
OTH. CORP. 12%
DOC. 5%
TRADE 11%
PROFESSIONAL 32%
52%
36%

CHEMICAL, METALLURGICAL AND OTHER ENGINEERING
N=177

EXT. 15%
LOCAL 19%
OTH. CORP. 22%
DOC. 11%
TRADE 11%
PROFESSIONAL 22%
56%
52%

MECHANICAL ENGINEERING
N=180

EXT. 13%
LOCAL 27%
PROF 13%
TRADE 15%
DOC. 7%
OTH. CORP. 25%
65%
59%

ELECTRICAL ENGINEERING
N=533

EXT. 9%
LOCAL 27%
PROF. 14%
TRADE 9%
DOC. 14%
OTH. CORP. 27%
63%
68%

of a need for certain information until he actually encounters it. Recognition of the need was stimulated by the acquisition of the information — rather than the reverse — in one-sixth of the instances reported. This situation can occur when information is pointed out spontaneously by someone else. It seems to account for a slight majority of instances of that sort, regardless of the respondent's technical field. It may happen also when the respondent is seeking information for development of his general competence; information that itself stimulated recognition of a need was encountered in 35% to 40% of those instances reported.

FIGURE 3.3

CIRCUMSTANCES OF ACQUISITION

	TOTAL N=1852	SCIENTISTS N=654	ENGINEERS N=966
SPECIFICALLY SOUGHT	47%	42%	53%
POINTED OUT	32%	33%	30%
GENERAL COMPETENCE	21%	25%	17%

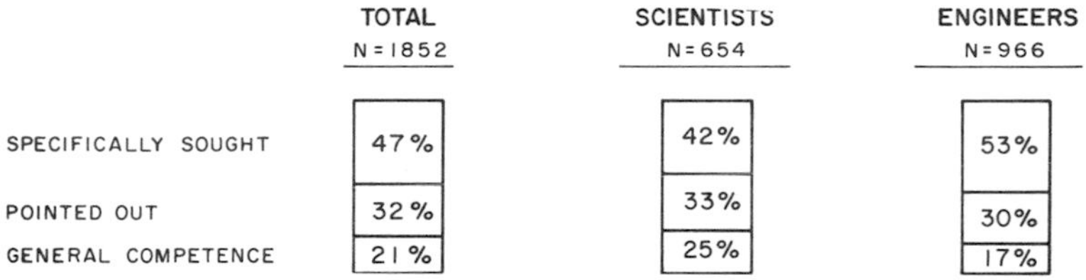

As Figure 3.3 indicates, there are no striking differences between scientists and engineers in the reported incidence of these circumstances. Scientists do, however, report a greater fraction of cases in which the information they used was acquired originally as a consequence of activities directed toward general competence, rather than a specific task.

After allowing for differences between professional groups, are differences in the circumstances of acquisition associated with differences in the use of sources of information? The answer is yes. For competence-oriented, as opposed to problem-oriented, quests for information, sources outside the corporation are used substantially more often. As would be expected, documentary sources are most common in these circumstances. The detailed data comparing problem-oriented and competence-oriented communication are given in Figures 3.4 and 3.5, in which results are reported separately for scientists and engineers.

FIGURE 3.4

PROBLEM-ORIENTED COMMUNICATION

(A) *Specific Search*

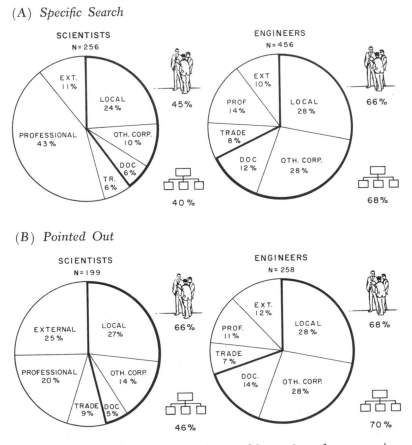

(B) *Pointed Out*

Figure 3.4, on the sources used in problem-oriented communication, throws some light on the differences between scientists and engineers. For situations in which the information was sought directly or was pointed out, local interpersonal communication is equally common for both groups. Scientists or engineers employed in the respondent's own establishment provided the substance of the information through interpersonal communication in slightly more than one-quarter of the reported instances both for scientists

and for engineers. The marked differences which we found between the two professional groups appear in their use of other
sources. Engineers report a greater incidence of interpersonal communication with people in other parts of their own corporations,
whereas a larger fraction of interpersonal communication by scientists is with individuals employed outside their own corporations. When using documents, engineers tend to consult corporate
reports or trade publications, while scientists make greater use of
the professional literature.

"Competence-oriented" communication implies such activities
as browsing, reviewing new developments, brushing up in a field,
developing familiarity with a new field, and so forth.[4] In these
circumstances, greater differences are apparent between scientists
and engineers (Figure 3.5). As would be expected, documentary

FIGURE 3.5

COMPETENCE-ORIENTED COMMUNICATION

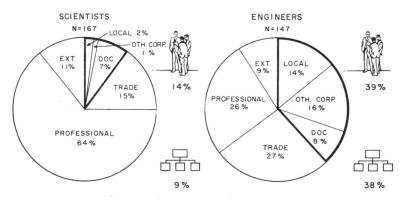

sources are dominant, even for engineers. Among scientists they
are almost entirely made up of articles and books drawn from
the "professional" literature; among engineers the use of profes-

---

[4] In discussion of the intended or actual function of information acquired,
it is common to distinguish between these "competence" functions and "ex-
haustive search" of some field. The latter is aggregated with other competence-
oriented modes in our data. Separate analysis disclosed no differences between
the pattern of sources in the two cases.

sional sources is matched by the use of information from "trade" sources (i.e., trade magazines, commercial catalogues and technical publications, and technical reports of other organizations). Engineers make far greater use of interpersonal communication in these circumstances. In virtually all cases in which a scientist reports competence-oriented interpersonal communication, the source is someone outside his own organization.

THE CONTEXT OF USE

We turn now to an analysis of transfer in terms of the three remaining variables in the situation: the organization, the task, and the user, himself. In this section we consider these three factors separately; in the next we shall use multivariate techniques which take them into account simultaneously. As before, we will control for the differences between scientists and engineers in most of the comparisons.

The Organizational Context

The principal distinction as to the organizational setting of work is between central laboratories and units which are closely linked, both geographically and organizationally, to operating divisions in the same corporation. Figure 3.6 summarizes data on the sources of information by organization, grouping them in these two main categories.[5] As this indicates, engineers and scientists in the central laboratories used sources outside their own corporation in approximately two-thirds of the instances they described. That situation

---

[5] The reader will recall that returns from 13 establishments in these corporate surveys were grouped for analysis into nine "businesses." The BASIC, POLY, and CONTROLS businesses include returns from more than one establishment doing similar work, usually with close geographic relation, within the same corporation. Data from the three establishments of POLY are grouped here with the operating divisions. These relatively small laboratories, although close geographically and organizationally to manufacturing and marketing organizations, may have a slightly different relationship with those organizations from that of the other groups we class as "operating" divisions. Note also that in Figure 3.6 we have not classified the Advanced Development Division of the DATA Corporation which fits neither category.

FIGURE 3.6: INFORMATION SOURCES BY BUSINESS

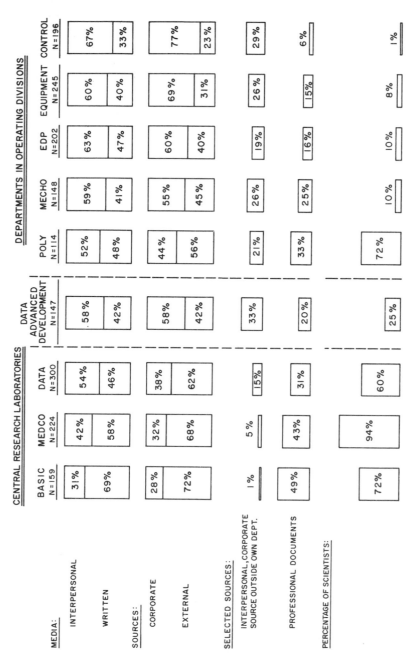

FIGURE 3.7

Sources by Work Setting

(*Chemists only*)

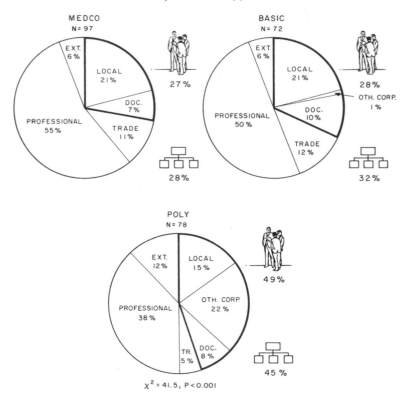

$\chi^2 = 41.5, P < 0.001$

is reversed in a typical operating division, where from three-fifths to three-quarters of the responses cite sources within the same firm. One part of the difference between these organizations seems to lie in the much greater incidence of interpersonal communication with men who are employed in the same firm but located outside of the respondent's own technical department. That type of inter-personal information transfer occurs in from one-fifth to one-quarter of the cases within operating divisions and in a much smaller, even negligible, proportion of cases in the central laboratories.

FIGURE 3.8

Sources by Work Setting
(Electrical Engineers in DATA Corporation)

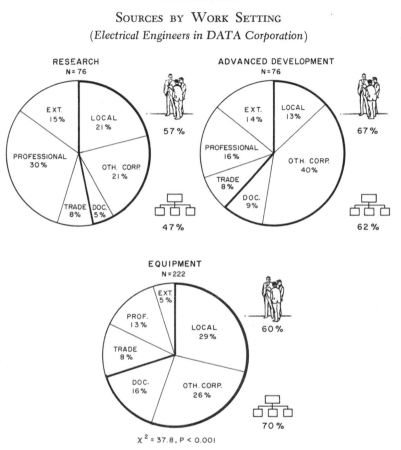

$\chi^2 = 37.8, P < 0.001$

These differences in communications patterns, which we are ascribing to differences between organization types, might merely reflect different employment patterns or the fact that the work of central laboratories is primarily scientific, while that of the operating divisions is primarily engineering. The same differences are found, however, when we control for professional field. Figure 3.7, for example, shows that there is a marked difference between the responses from chemists in research laboratories and those in operating divisions. The chemists in POLY report a significantly greater use of sources within their own corporation than do the others.[6] That difference is especially strong when one considers interpersonal communication with other sources in the corporation: 22% in POLY and less than 1% in the two central laboratories.[7]

As another example, the summary in Figure 3.8 of responses from electrical engineers in the DATA Corporation is limited to respondents in the same professional field, employed in the same corporation, and working in the same product field. The same pattern emerges: A majority of instances reported by engineers in the Research Laboratory involved sources outside the DATA Corporation, while three-quarters of the responses from those in the Equipment Division cited internal sources. The incidence of use of professional documents is twice as large among electrical engineers in the research lab as it is in the other two divisions. In these data we also see that the use of interpersonal communication with other corporate sources is substantially greater in the linking organization, Advanced Development, than in the other two divisions. Of course, an effective organization in its position should be able to draw on both the fundamental work done in the research laboratories and the knowledge of evolving operating needs originating in the product divisions. Hence a high incidence of interdivisional communication is consistent with, and probably a reflection of, the basic mission of the organization.

Characteristics of the Task

Two of the factors discussed above can be interpreted as indirect indicators of task content; they are the individual's professional field and his place of employment. We turn now to a third and more particular descriptor: a characterization of the specific task to which the information in question was applied. Table 3-1 presents the set of statements used in the questionnaire to describe

---

[6] This is not because that organization employs junior chemists; 40% have Ph.D. degrees.

[7] There is more than geography at work here. The MEDCO laboratory is at the site of a major manufacturing facility and near corporate administrative and marketing staff departments. The term "central laboratory" describes a state of mind or sense of mission rather than a geographic location.

T<small>ABLE</small> 3-1

Task Categories in Relation to Original Questions

Each respondent was asked to check one of several descriptions to complete the sentence: "The information was applied to a task which could be characterized as an example of the . . ."

(A) The characterizations offered on the IEEE questionnaire and (with the exception of item g) in MECHO and CONTROL are listed here with the category used in analysis:

a. formulation or testing of scientific theories, concepts, or models	Research
b. empirical scientific investigation of physical phenomena	Research
c. formulation, development, and investigation of new approaches to technical objectives	Development
d. combination and integration of generally available designs and components into desired products, processes, and test procedures	Design
e. refinement of existing products, processes, or test procedures	Design
f. conduct of tests of materials, products, or processes	Test
g. formulation, programming, or testing of approaches to data-processing problems	Development
h. following (please describe): _____	

(B) The following items were used in EDP and all of the DATA Corp.:

a. formulation or testing of scientific theories, concepts, and/or models	Research
b. formulation or analysis of a logical operation or a data-processing language	Research
c. empirical investigation of physical phenomena	Research
d. formulation, development, or investigation of new approaches to problems to which data processing will be applied	Development
e. formulation, development, or investigation of new approaches to the design of data-processing equipment or components	Development
f. programming and/or coding for a particular problem	Design

TABLE 3-1 (continued)

 g. combination, integration or refinement of designs and components into desired equipment systems, components, or test procedures .... Design

 h. conduct of tests of equipment, components, or materials .... Test

 i. testing or debugging of a data-processing program .... Test

 j. following (please describe): _____

(C) In the "materials" version of the questionnaire used in MEDCO, BASIC, and POLY, these items were used:

 a. formulation or testing of scientific theories, concepts, or models .... Research

 b. empirical investigation of physical phenomena Research

 c. synthesis of a compound or material .... Development

 d. determination of the properties of a compound or material .... Analysis

 e. analysis of a compound or material .... Analysis

 f. testing of a material or compound for a specific application .... Test

 g. design or modification of a manufacturing or pilot plant process .... Design

 h. construction of a physical model or the fabrication of a piece of equipment .... Design

 i. following (please describe): _____

the task in which the information was employed. With each descriptive phrase we indicate in the table which of five summary categories — research, development, design, test, or analysis — we used to characterize a specific response.

Consistent differences in the means of acquiring information are associated with differences in task, both in central research laboratories and in operating divisions (see Figure 3.9). In both work settings information for research tasks was gained from external sources more often than was information for design and development tasks, while interpersonal communication within the corporation was used less often. The data suggest also that analytic and testing work in the operating divisions is of a different character from that in the research laboratories; in the latter very heavy use

FIGURE 3.9

INFORMATION SOURCES BY TASK

(A) Central Laboratories

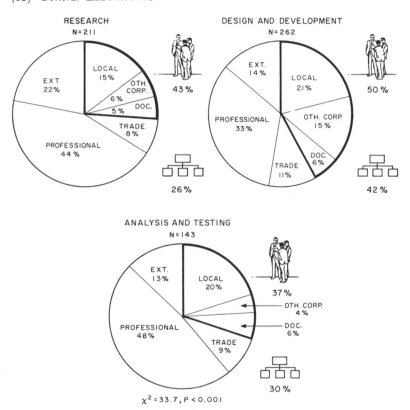

$$\chi^2 = 33.7, P < 0.001$$

is made of professional documents, while the most common source in the operating divisions is interpersonal communication within the firm.

Although these differences still hold when we control for profession, the data here are not so persuasive. Scientists in central research laboratories do use external sources of information significantly more: 77% for research, as opposed to 66% for design and development. Engineers in the operating division, typically performing very few research tasks, do use external sources of in-

FIGURE 3.9 (continued)

(B) Operating Divisions

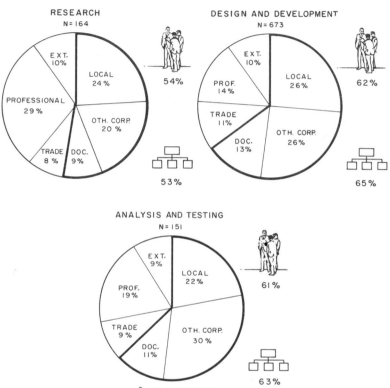

$\chi^2 = 27.4, P < 0.005$

formation slightly more often for those tasks and do use professional documents significantly more often.

The Personal Context of Use

Our background information on the respondent includes several different measures, which we group in two categories: experience and personal commitment.

Experience: We considered two dimensions of an individual's experience with potential sources of information: (1) an intellectual aspect, his experience in the technical discipline from which the

FIGURE 3.9 (continued)

(C) Selected Sources

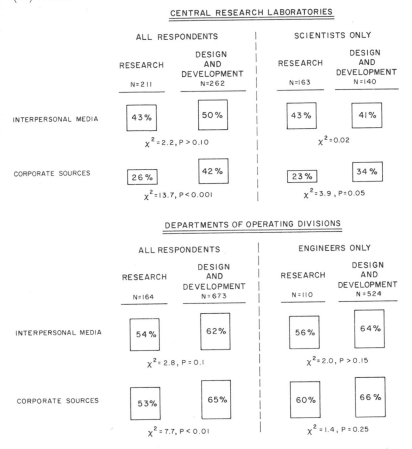

CENTRAL RESEARCH LABORATORIES

ALL RESPONDENTS | SCIENTISTS ONLY

RESEARCH N=211 / DESIGN AND DEVELOPMENT N=262 | RESEARCH N=163 / DESIGN AND DEVELOPMENT N=140

INTERPERSONAL MEDIA: 43% / 50%, $x^2 = 2.2, P > 0.10$ | 43% / 41%, $x^2 = 0.02$

CORPORATE SOURCES: 26% / 42%, $x^2 = 13.7, P < 0.001$ | 23% / 34%, $x^2 = 3.9, P = 0.05$

DEPARTMENTS OF OPERATING DIVISIONS

ALL RESPONDENTS | ENGINEERS ONLY

RESEARCH N=164 / DESIGN AND DEVELOPMENT N=673 | RESEARCH N=110 / DESIGN AND DEVELOPMENT N=524

INTERPERSONAL MEDIA: 54% / 62%, $x^2 = 2.8, P = 0.1$ | 56% / 64%, $x^2 = 2.0, P > 0.15$

CORPORATE SOURCES: 53% / 65%, $x^2 = 7.7, P < 0.01$ | 60% / 66%, $x^2 = 1.4, P = 0.25$

information was obtained, and (2) an organizational aspect, his seniority in his corporation.

Questions 8 and 9 asked the respondent to identify the specific field from which the information came, the class of problem or application for which it was useful, and whether or not he considered himself experienced in the relevant discipline and application. We reasoned that a man faced with learning something about an unfamiliar technical problem has to face a procedural problem first. He must find out where to look. In that case he is likely to rely on some person whom he considers to be experienced in the field in question. Our data show that interpersonal communication and local sources were used significantly more frequently when the respondent considered himself inexperienced in the source discipline. The data are shown separately in Figure 3.10 for operating divisions and central research laboratories. We can observe the same effect if we control for profession as well as for

FIGURE 3.10

Sources by Experience in Discipline

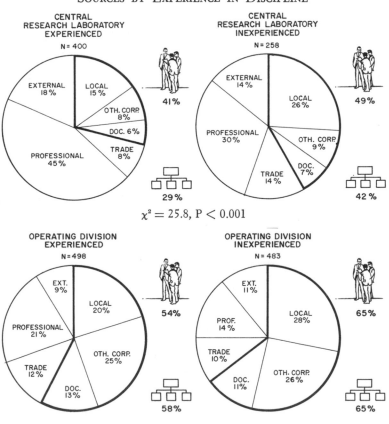

locale. No consistent pattern emerges, however, in relation to the measure of application experience.

Our measure of organizational experience is the number of years the respondent has been employed by the firm for which he currently works. The data again support our prior conjecture and the finding of our exploratory study. Although the effect is not large, there is a definite positive association between the incidence of use of sources within the firm and the respondents' seniority. Men with more than 10 years' tenure, as shown in Figure 3.11, use corporate sources outside their own establishment 1½ times

FIGURE 3.11

Sources by Degree of Seniority

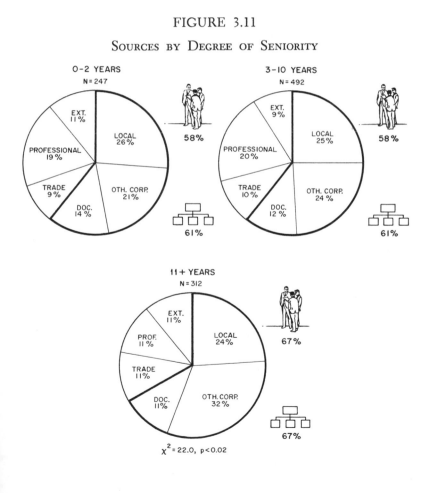

$$x^2 = 22.0, \ p < 0.02$$

more frequently and professional publications only half as often as men with brief tenure.

Personal Commitment: There are important individual differences in the orientation toward work on the part of engineers and scientists in R&D. Variations in the degree of commitment to the technical content of work were of particular interest to us. Although there is no direct measure of this sort of commitment, we

FIGURE 3.12

SOURCES BY EDUCATIONAL LEVEL

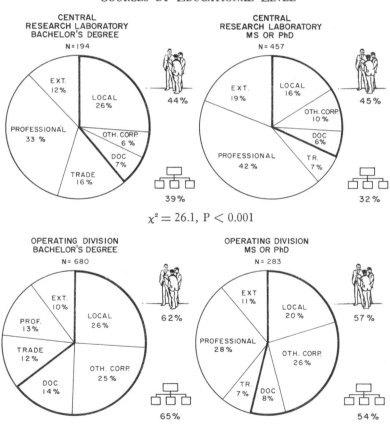

$\chi^2 = 26.1$, P $<$ 0.001

$\chi^2 = 36.7$, P $<$ 0.001

can construct an index from objective data. We discuss this phenomenon more completely in Chapter 4, in which we define such an index. Right now we will consider, as separate factors, several manifestations of a man's commitment to the development of professional competence.

A man with a high degree of commitment tends to pursue his formal education further and to make education a continuing process — for example, by joining professional societies, attending their meetings, and reading a large number of technical periodicals. These men tend to be more productive in professional terms, i.e., in the publication of papers, although this does not imply either greater or lesser productivity for their employers' purposes. We

FIGURE 3.13

Sources by Professional Activity

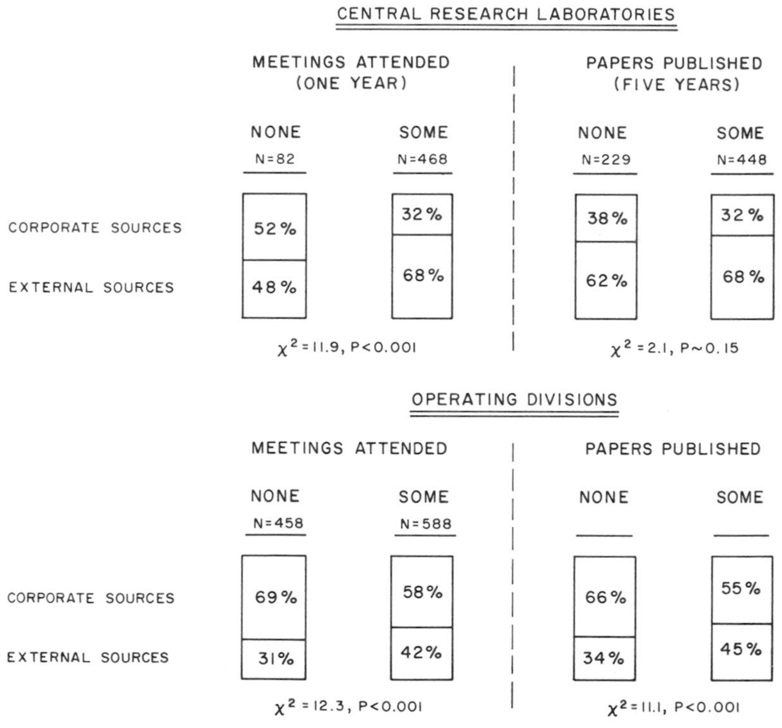

CENTRAL RESEARCH LABORATORIES

	MEETINGS ATTENDED (ONE YEAR)		PAPERS PUBLISHED (FIVE YEARS)	
	NONE N=82	SOME N=468	NONE N=229	SOME N=448
CORPORATE SOURCES	52%	32%	38%	32%
EXTERNAL SOURCES	48%	68%	62%	68%
	$\chi^2 = 11.9, P < 0.001$		$\chi^2 = 2.1, P \sim 0.15$	

OPERATING DIVISIONS

	MEETINGS ATTENDED		PAPERS PUBLISHED	
	NONE N=458	SOME N=588	NONE	SOME
CORPORATE SOURCES	69%	58%	66%	55%
EXTERNAL SOURCES	31%	42%	34%	45%
	$\chi^2 = 12.3, P < 0.001$		$\chi^2 = 11.1, P < 0.001$	

would expect that engineers and scientists with such commitment
to maintaining competence would use professional publications
more often, and local sources of information less often, than the
average respondent.

As Figure 3.12 shows, the anticipated association appears quite
clearly in relation to educational level, both in central laboratories

FIGURE 3.14

SELECTED SOURCES BY PROFESSIONAL ACTIVITY

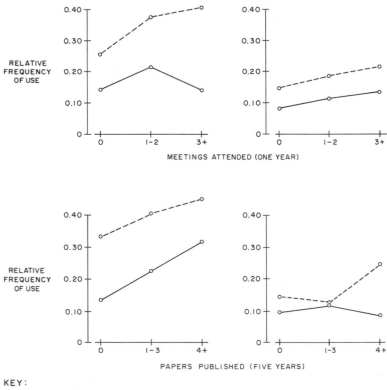

KEY:

——— OPERATING DIVISIONS
‒ ‒ ‒ ‒ CENTRAL RESEARCH LABORATORIES

FIGURE 3.15

Sources by Periodical Readership

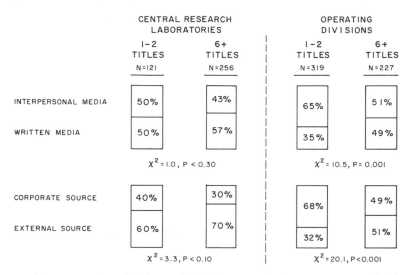

	CENTRAL RESEARCH LABORATORIES		OPERATING DIVISIONS	
	1-2 TITLES N=121	6+ TITLES N=256	1-2 TITLES N=319	6+ TITLES N=227
INTERPERSONAL MEDIA	50%	43%	65%	51%
WRITTEN MEDIA	50%	57%	35%	49%
	$x^2 = 1.0$, $P < 0.30$		$x^2 = 10.5$, $P = 0.001$	
CORPORATE SOURCE	40%	30%	68%	49%
EXTERNAL SOURCE	60%	70%	32%	51%
	$x^2 = 3.3$, $P < 0.10$		$x^2 = 20.1$, $P < 0.001$	

and in operating divisions. A similar association appears, in both work settings, between use of sources within the company and two measures of professional activity: attendance at meetings of professional societies and authorship of published books or articles. This relationship is summarized in Figure 3.13. The principal variations appear in the use of professional publications and in interpersonal communication with persons outside the respondent's firm. Data on use of these selected sources, in relation to professional activity, are depicted in Figure 3.14.

Another measure of personal commitment is the number of periodicals regularly read.[8] As the reader will recall, one question asked the respondent to list the "scientific, technical, and trade periodicals" that he read "regularly for the purpose of keeping up on my field." We reviewed the lists to exclude nontechnical publications (e.g., *Business Week*) and then counted the number listed.

---

[8] The questionnaire also asked for the number of U.S. patents filed by the respondent in the last five years. No consistent relationship could be found in contingency tables tabulating this measure of technical productivity and the use of various information sources.

In both work settings respondents listing six or more periodicals made less frequent use both of corporate sources and of interpersonal communication than did those listing only one or two titles (Figure 3.15).

The data summarized in Figures 3.6 through 3.15 suggest that the propensity of an individual to use information from sources outside his firm is not independent of his task or the organizational milieu, nor of his own professional affiliation, experience, or commitment to development of professional competence. Two questions occur naturally: (1) If these several different aspects of the circumstances of work were considered simultaneously, would all continue to appear relevant to understanding the use of alternative sources of information? and (2) How strong is the association between these characteristics and the varying use of sources of information, specifically, what proportion of the variance of the latter is explained by the former?

The foregoing figures are based on contingency tables. The use of this method of analysis becomes awkward very rapidly as the number of variables increases; other multivariate techniques, such as regression analysis, are more appropriate for the questions just posed. Although they are not conventionally applied to the analysis of data of this type, we used both multiple linear regression and linear discriminant techniques to analyze subpopulations in our sample. The results confirm our expectation that there is a strong association — explaining nearly half the variance — between the use of alternative sources and these several circumstances of work. The analysis shows, further, that a predictive model continues to be improved by the addition of variables until all of the main categories are represented. The remainder of this chapter describes the formulation of these predictive models and the results of their application.

The Group as a Unit of Analysis

Throughout most of this study the "instance" reported by a single respondent is used as the unit of analysis. For regression analysis, we shall move to the next higher level of aggregation (Level 2 in Table 2-1) and use the "group of immediate colleagues" (henceforth termed "group") as the basic unit. The reader will recall that the subset of instances reported by members of a group is considered to be a sample from the total set of all instances of information acquired by that group. Hence, if in 55% of the instances described by members of a group the source of the information was an employee of the same firm, we take the fraction .55 as an estimate of the relative frequency of use of that type of source by that group. This statistic — the fraction of instances in which members of a group used sources within their own company — is treated as the dependent variable.

The smallest formal unit of organization in each establishment (usually a "section") defined the bounds of the unit of analysis. The companies' organization charts were used to assign respondents to particular groups. There were 86 groups in 9 establishments in the MEDCO, BASIC, and DELTA companies for which we had usable responses from 4 or more persons.[9]

Scores for each group were computed on a scale from zero to one for each of 14 variables. A group's score on the dependent variable was the fraction of members of that group reporting instances in which they used sources within their own firm. Of the 13 independent variables, 11 were based on the factors used in the preceding analysis, e.g., task, profession, experience, education, professional activity. Two other measures, the prior circumstance of acquisition and the time interval since the date of the instance, were included as controls.[10] The nature of each independent variable was defined by dividing the possible individual answers into

---

[9] We were unable to associate respondents with groups in any part of the DATA Corporation and in the larger of the two establishments in the BASIC Corporation. The 778 individuals assigned to the 86 identifiable groups constitute 76% of the respondents in the remaining 9 establishments.

[10] Both competence-oriented communication and a lengthy interval since the date of the instance were shown here to be associated with a greater incidence of use of written sources, and hence greater use of sources originating outside the firm. They are included as controls to avoid the spurious associations that might arise from their being correlated with the other 11 variables. The results showed that it is useful to control for the prior circumstances but not for interval.

two sets and assigning a zero designation to answers in one set
and a one designation to the others. The group score on each
variable was computed as the sum of the individual scores divided
by the number of members of the group who gave usable answers
on that question. The first variable, for example, defines the relative frequency within the group of the characteristic "experienced
in the discipline from which the information was obtained." In a
group of nine men, in which six answered "yes" to question 8, two
answered "no," and one gave no answer, the group score on variable 1, "discipline experience," would be $(6 \div 8) = .75$. Table 3-2
defines the 14 group variables and gives the means and variances
of the scores for 86 groups.[11]

Will knowledge of a group's score on any of the independent
variables substantially improve our ability to predict the incidence
of that group's use of sources within the same company? If so, to
what extent?

For all 86 groups, 45% of the variance in the dependent variable,
which measures the relative use of sources within the same company, can be explained by linear regression on 5 variables: 4 de-
scriptors plus the control for competence mode.[12] The model and
results of the analysis are given in Table 3-3. The measures used
are indicative of professional affiliation (X_3), commitment to competence (X_{12}), task (X_{11}), and experience (X_5).[13] For predictive
purposes, the first two are given greatest weight, with the other
variables carrying approximately equal smaller weights. Stepwise
regression showed that no important improvement in prediction
could be achieved by adding any of the other independent variables. Further notes on this analysis are given in Appendix D.

---

[11] The mean score of .513 for the dependent variable (company sources)
compares with the overall incidence of .52 for company sources reported by
all respondents in the 1965 corporate surveys.

[12] These results may be the product of chance, rather than a reflection of
systematic relationships within the population under study. Since we cannot
claim to have a random sample from a multivariate normal population, valid
probability statements cannot be made. If that requirement were met, however, the F ratio of 13.18 for the results in Table 3-3 would indicate a very
low probability of chance causes.

[13] Almost all of these groups worked in operating divisions, thus excluding
the possibility of using the organizational context as a variable in this analysis.

TABLE 3-2: **Variables and Score Categories for Group Analysis**

Variable	Description	Answers Scored as "ones" *	Scores for 86 Groups		
			Mean	Variance	Correlation with dependent variable
Dependent	Company Source	Local and other corporate interpersonal; company documents	.513	.0628	1.0
X_1	Discipline Experience	Yes	.535	.0440	−.11
X_2	Application Experience	Yes	.712	.0292	−.08
X_3	Scientific Profession	Any scientific field on Question 20	.367	.1812	−.51
X_4	Low Seniority	0–2 years	.235	.0273	−.06
X_5	High Seniority	11 or more years	.338	.0543	.06
X_6	Higher Education	M.S. or Ph.D. degrees received	.336	.0701	−.50
X_7	Meetings Attended	One or more in past year	.681	.0397	−.46
X_8	Papers Published	One or more in five years	.368	.0627	−.40
X_9	Patents Issued	One or more in five years	.318	.0682	−.21
X_{10}	Recent Instance	Within last three weeks	.806	.0195	.14
X_{11}	Scientific Task	Answer in "research" category on Question 10	.197	.0387	−.38
X_{12}	Many Periodicals Read	6 or more titles listed	.286	.0513	−.55
X_{13}	Competence Mode	Intent was to develop general competence	.134	.0138	−.38

* Other answers scored as zeros.

TABLE 3-3

Results of Regression Analysis for 86 Groups

Equation for estimating fraction of sources within the same company:
$\hat{Y} = -.147X_3$ (Scientific field) $+ .199X_5$ (Seniority) $-.208X_{11}$ (Scientific Task) $- .354X_{12}$ (Periodicals read) $- .419X_{13}$ (Competence Mode) $+ .698$

Multiple r $= .672$ Multiple $r^2 = .45$

	Beta Weights	Standard Errors of Beta Weights
X_3	$-.250$	.105
X_5	.185	.087
X_{11}	$-.163$	.092
X_{12}	$-.320$	.112
X_{13}	$-.196$	.091

Discriminant Analysis

As another experiment in predicting individual propensities to use different sources for information, we used linear discriminant analysis of data reported in the Equipment Division and the Research Laboratory of the DATA Corporation. These two organizations provide an interesting pair. One is an operating division; the other, a central laboratory. Both are located within the same corporation and are concerned with a common product area. As independent variables we chose the same characteristics used for the regression analysis on group scores, excluding the "control" scores for interval and mode, and then added a variable to identify work locale (i.e., the division in which the respondent was employed). In a sense, then, this replicates the foregoing analysis of the predictive power of these characteristics, using the same measures but applying a different statistical technique to a different body of data and using the individual rather than the group as the unit of analysis.

Discriminant analysis is a method of predicting the probability of membership in a specified categorical group by means of a linear combination of scores on specified dimensions. In our case we want to estimate the probability that a given respondent will describe as his most recent instance of information transfer one in which he used a source of a given type.

As was true for the regression analysis, to carry out the analysis we had to transform categorical answers into numerical values. Four of the dimensions represent dichotomous measures: There are two work settings and two professional fields, as well as yes or no answers for the two experience questions. These were readily scored zero or one. Although the other seven variables had five or six categories each, the categories have a natural order. In five cases in which the categories represent ranges along an underlying ratio scale, we scored them to provide a nonlinear transformation, on the assumption (supported by our contingency analysis) that the main effects would be observed at the lower end of the scale. The other two dimensions, education level and type of task, were simply scored in rank order. Table 3-4 summarizes the scale categories used for this analysis.

The first step of analysis was to try to discriminate between instances of the six categories of sources used throughout our analysis. The first six columns in Table 3-5 report the mean scores on the 11 dimensions for respondents grouped according to the sources from which they acquired the information in the specific case reported on the questionnaire. For each of the four dichotomous measures, the mean scores simply represent the percentage of respondents in each group reporting the characteristic assigned a score of one on that dimension. For example 40% of those reporting use of "professional" documents were engineers, and 78% were in the research laboratory. Mean scores on the other dimensions must be interpreted cautiously, since the scales are arbitrary; they do, however, indicate the central tendencies for respondents in each group. In total, complete data on these 11 dimensions were available for 329 respondents in the two divisions.

The results of a first attempt at discrimination between all six groups suggested that two underlying groupings could be distinguished by use of a single discriminant score. One group comprised the first three categories, corresponding to the "corporate source" category used for the preceding analysis; the second combined "professional" and "external-interpersonal" source categories. The "trade document" category, which is a residual aggregation of various other documents originating outside the firm, could not be systematically associated with either group.

TABLE 3-4

Scale Values for Discriminant Analysis

Number assigned:	0	1	2	3	4	5	6
Characteristic:							
Locale	Equipment	Research					
Discipline Experience	Yes	No					
Application Experience	Yes	No					
Profession	Scientist	Engineer					
Seniority (years)	—	0–1	1–2	3–5	6–10	11+	
Education	—	High School	Certificate	Received BS	Some graduate study	Received MS	Received Ph.D
Meetings (in last year)	—	None	1	2	3	4+	
Papers Published (five years)	—	None	1	2–3	4–6	7+	
Patents Filed (five years)	—	None	1	2–3	4–6	7–15	16+
Periodicals Read	—	None	1	2	3–5	6+	
Task	—	Fundamental Research	Empirical Research	Development	Design		

TABLE 3-5

Linear Discriminant Analysis

(329 Respondents)

DATA Corporation, Equipment and Research Divisions

(Group Mean Scores)

Variables	Notes on Scoring	Interpersonal		Company Documents	Trade	External-Interpersonal	Professional	Scaled Discriminant Weights		Coefficients for Estimating Equation
		Local	Other Corporate					6 Groups	2 Groups	2 Groups
Discipline Exp.	1 = NO	.54	.45	.52	.52	.35	.28	−.29	−.29	−.50
Application Exp.	1 = NO	.30	.21	.35	.29	.16	.29	.21	.20	.38
Profession	0 = Scientist									
	1 = Engineer	.70	.80	.28	.84	.58	.40	−.03	−.14	−.25
Seniority		3.28	3.45	3.74	3.52	3.19	2.91	−.27	−.28	−.20
Education	3 = BS									
	5 = MS	4.23	4.26	3.74	4.55	5.37	5.46	.34	.34	.24
Meetings	2 = ONE	1.97	1.86	2.06	2.52	3.19	2.98	.25	.33	.24
Papers	2 = ONE	1.93	2.05	1.71	2.16	3.56	3.52	.51	.40	.24
Patents	2 = ONE	1.63	1.68	1.74	1.81	1.98	1.57	−.17	−.08	−.06
Periodicals	4 = 3–5	3.56	3.47	3.65	4.10	4.16	4.15	.15	.19	.14
Task	1,2 = Research 3 = Development 4 = Design	2.94	3.03	2.94	2.77	2.14	1.95	−.46	−.49	−.39
Division	0 = Equipment 1 = Research	0.35	0.38	0.29	0.55	0.89	0.78	.26	.30	.29

The coefficients of the discriminant equation for two groups —
"company" sources and "external-interpersonal" plus "professional"
— are given in the last column in Table 3-5. Next to it appear
the weights adjusted to account for scale differences (given both
for the six-group and the two-group analysis to show that the best
discriminant function is approximately the same in either case).
On the basis of this discriminant score, 78% of the respondents
could be assigned to the correct group, as between those reporting
the use of sources within the DATA Corporation and those reporting outside sources. Table 3-6 gives a frequency count of respondents in the two groups according to eight intervals on the
scale of discriminant scores, an approximate measure of the probability of membership in each group at eight points on the scale,
and a table of hits and misses.

This analysis shows that, at least within these two large organizations, identifying the relevant task and organizational setting
plus a relatively small number of personal characteristics of the
individual concerned should make it possible to predict accurately
in nearly four out of five cases whether the information will be
acquired from within the firm or instead from a professional document or personal source outside the organization.

Results of IEEE Surveys

The IEEE respondents are a distinctive group, differing from
respondents in the corporate surveys both in personal and situational characteristics and in their relative use of sources for information transfer. Table 3-7 shows the reported incidence of use of
various sources and of various circumstances for acquisition of
technical information. The corresponding data for the respondents
in the corporate surveys, in total and for the most comparable
sub-group — electrical engineers — are restated for comparison.[14]

---

[14] In the case of interpersonal communication within the corporation, the
distinction made in the corporate surveys, between informants who were professionals employed within the same department and other informants in the
organization, could not be applied consistently to the much more varied organizational circumstances of the IEEE respondents. Hence, Tables 3-7 through
3-13 aggregate all instances of interpersonal communication within the respondent's own organization.

TABLE 3-6

Results of Discriminant Analysis

(A) Frequency distribution by discriminant score and type of source reported

	Respondents reporting	
Score	*Source within DATA Corp.*	*External-Interpersonal or Professional Document*
0.0– .99	8	1
1.0–1.99	44	2
2.0–2.99	47	8
3.0–3.99	33	16
4.0–4.99	23	34
5.0–5.99	10	48
≥ 6.0	3	21
	168	130

(B) Approximate probability of group membership

	Discriminant Score						
	1.0	2.0	3.0	4.0	5.0	6.0	7.0
Group:							
Source within DATA Corp.	.99	.94	.79	.51	.25	.11	.04
External-Interpersonal or Professional	.01	.06	.21	.49	.75	.89	.96

(C) Table of hits and misses

	Predicted * Source	
Actual Source	*Inside DATA Corp.*	*External-Interpersonal and Professional*
DATA Corp.	132	36
External-Interpersonal or Professional (Hits = 78%)	30	100

* Predictions were made by computing the discriminant score for each respondent and assigning him to the group with the higher indicated probability (see B above). Predictions were then sorted according to the actual source employed to produce the table above.

TABLE 3-7

Comparison of Results for IEEE and Corporate Surveys

| | Corporate Surveys | | IEEE Surveys |
	Total	*Electrical Engineers*	*Total*
Sources	(N = 1735)	(N = 533)	(N = 1075)
Organizational Inter-			
personal	42%	54%	17%
Organizational Documents	10	14	6
Trade Documents	10	9	25
Professional Documents	25	14	25
External-Interpersonal	13	9	27
Interpersonal Media	55%	63%	44%
Organizational Sources	52%	68%	23%
Circumstances of Acquisi-			
tion:	(N = 1852)	(N = 580)	(N = 1133)
Specific Search	47%	52%	35%
Pointed Out	32%	30%	31%
General Competence	21%	18%	34%

Members of IEEE Groups, in comparison to our other sample of Electrical Engineers, make considerably more frequent use of sources outside their own organizations. The increase is substantial in all three categories of external sources, namely, interpersonal communication, trade literature, and professional documents. The overall incidence of interpersonal communication is substantial (44%), although less than among the corporate respondents, and is weighted heavily toward informants outside the organization, in sharp contrast to the corporate case. Comparison of results from the several IEEE Groups (in Table 3-8) reveals some between-group differences, but the IEEE Group members, whatever their field, are more like each other than like the average electrical engineer in the other surveys.

To some extent, these differences in information transfer may be explainable by differences in characteristics of the respondents' background, work, or organizational setting. Overall, the IEEE respondents are more highly educated, more active professionally, and have slightly more seniority than their counterparts in our

TABLE 3-8

Comparison of Results Among IEEE Groups*

	Power Group	Computers' Group	Circuit Theory Group	IECI Group	Multiple Groups
Sources:	(N = 286)	(N = 337)	(N = 128)	(N = 47)	(N = 236)
Organizational Interpersonal	13%	22%	14%	28%	13%
Organizational Documents	4	9	6	2	4
Trade Documents	27	25	20	21	27
Professional Documents	22	18	44	13	33
External-Interpersonal	34	26	16	36	23
Interpersonal Media	47%	48%	30%	64%	36%
Organizational Sources	17%	31%	20%	30%	17%
Circumstances of Acquisition:	(N = 318)	(N = 356)	(N = 147)	(N = 50)	(N = 247)
Specifically Sought	37%	31%	39%	44%	34%
Pointed Out	34%	35%	24%	32%	25%
General Competence	29%	34%	37%	24%	41%

* The first four columns tabulate data from respondents affiliated with only one of the four selected groups, although they may have had other group affiliations as well. The "Multiple Group" category tabulates those belonging to two or more of the selected groups.

TABLE 3-9

Comparison of Respondents' Characteristics

	Electrical Engineers in Surveys in Corporations	*IEEE Groups*
Seniority in present organization: 0–2 years	23%	25%
3–10 years	49%	42%
11 + years	28%	33%
Highest Education Master's Degree	19%	31%
Ph.D. Degree	6%	7%
One or more patent applications filed in previous five years	33%	29%
One or more publications in previous five years	33%	41%
Professional meetings attended in the previous year: at least one	57%	80%
three or more	14%	40%
Above average score on index of commitment to the development of competence*	33%	71%
Type of task to which the information was applied: Research	13%	15%
Development	39%	32%
Design	43%	43%
Analysis & Test	5%	10%
Number of subjects	594	1195

* Among all engineers and scientists responding to the surveys in corporations, this index (defined in Chapter 4) had an average score of approximately 3.0.

other surveys. (See Table 3-9 and Appendix C). These differences are to be expected from the nature of the IEEE Group as an institution. Only a few respondents in the corporate surveys held supervisory positions, while one-half of the IEEE respondents were project heads, group leaders, supervisors, or managers. The IEEE

TABLE 3-10

Sources of Information for Selected Tasks
(*Individual Contributors Only*)

	Tasks		
	Research	*Development*	*Design*
Sources	(N = 48)	(N = 126)	(N = 142)
Organizational Interpersonal	6%	16%	21%
Organizational Documents	2	10	5
Trade Documents	21	21	32
Professional Documents	52	34	24
External-Interpersonal	19	19	18
	(Chi Square = 22.8, p < .001)		
Interpersonal Media	25%	35%	39%
Organizational Sources	8%	26%	26%

responses come, furthermore, from a more heterogeneous set of organizational settings: one in five is from an establishment of the government, a university, or a nonprofit corporation. Those within the industrial sector represent a greater range and diversity of product fields than is found in the four corporations we studied.

At the same time, the situations represented by the two groups of respondents are similar along certain dimensions. The content of the relevant tasks, as measured both by the definition of the technical field and by our scale for "type of task," is comparable.

TABLE 3-11

Sources by Job Rank, Within Operating Departments

	Individual Contributors	Supervisors
Sources	N = 131	N = 260
Organizational Interpersonal	16%	15%
Organizational Documents	8	3
Trade Documents	27	30
Professional Documents	27	18
External-Interpersonal	22	34
Interpersonal Media	38%	49%
Organizational Sources	24%	18%

TABLE 3-12

Sources of Information in Various Work Settings

(Controlled on Organizational Position)

Work Setting

(A) Individual Contributors

Sources	Education and Nonprofit Laboratories	Other Research Laboratories	Government and Industrial Development Laboratories	Industrial Operating Departments	Government Operating Departments
	N = 43	N = 26	N = 62	N = 131	N = 60
Organizational Interpersonal	14%	12%	26%	16%	15%
Organizational Documents	—	8	11	8	12
Trade Documents	26	23	21	27	18
Professional Documents	32	42	29	27	37
External-Interpersonal	28	15	13	22	18
Interpersonal Media	42%	27%	39%	38%	33%
Organizational Sources	14%	20%	37%	24%	27%

TABLE 3-12 (continued)

(B) *Supervisors*

Work Setting

Sources	Education and Nonprofit Laboratories	Other Research Laboratories	Industrial Development Laboratories	Government Development Laboratories	Industrial Operating Departments	Government Operating Departments
	N = 49	N = 38	N = 58	N = 31	N = 260	N = 47
Organizational Interpersonal	14%	18%	34%	19%	15%	21%
Organizational Documents	6	10	9	7	3	4
Trade Documents	22	24	14	19	30	28
Professional Documents	27	32	15	29	18	21
External-Interpersonal	31	16	28	26	34	26
Interpersonal Media	45%	34%	62%	45%	49%	47%
Organizational Sources	20%	28%	43%	26%	18%	25%

The IEEE Groups were selected to be comparable to the work of the Electrical Engineers in our sample; the respondents' descriptions of tasks are similarly distributed for the two groups (see Table 3-9). Furthermore, in general terms the professional contexts within which these groups were trained and now conduct their work can be assumed to have been the same.

In summary, when we compare these two groups of respondents, those Electrical Engineers in the surveys within corporations and those in IEEE Groups, the two are substantially different with respect to individual characteristics of respondents and the organizational context within which they work, but are more comparable in the technical content of work and its professional context.

Earlier in this chapter we presented evidence of systematic relationships between the use of sources for information transfer and those characteristics of technical work, of technical workers, and of the organizational settings in which they are found. The IEEE data show the same relationships, but to a significantly lesser degree.

Let us consider some of the detailed results. The type of task to which the information was applied is a characteristic which is similarly distributed in both the IEEE and corporate data. Table 3-10 shows that the association with information sources is similar, among IEEE respondents, to that observed in the corporate data, i.e., lesser use of sources within the organization when the task is "research." Although sources outside the organization appear with equal frequency for both development and design tasks, the relative usage of trade and professional documents is reversed between the two cases. Development tasks, with their higher relevance to "professional" contributions, are associated with a greater use of professional documents.

Supervisors constitute half of the set of IEEE respondents. For work done within operating departments, there is a tendency for supervisors to use outside sources slightly less than do individual contributors. Supervisors, furthermore, refer less often to professional documents, and more often to interpersonal media, among those external sources. (See Table 3-11.) Table 3-12 shows that the use of information sources also varies among the organizational settings (more diverse here than in the corporate surveys), for

TABLE 3-13

Sources of Information in Laboratories, by Group Membership*

(Controlled on Organizational Position)

Sources	Individual Contributors		Supervisors		
	Circuit Theory N = 26	Computers N = 44	Circuit Theory N = 22	Computers N = 47	Multiple Groups N = 64
Organizational Interpersonal	12%	32%	27%	30%	19%
Organizational Documents	12	7	0	21	3
Trade Documents	8	30	23	15	25
Professional Documents	65	21	36	11	22
External-Interpersonal	4	11	14	23	31
Interpersonal Media	16%	43%	41%	53%	50%
Organizational Sources	24%	39%	27%	51%	22%

* Laboratories include both Research and Development Organizations.

TABLE 3-14

Sources of Information in Operating Departments, by Group
(Controlled on Organizational Position)

Sources	Individual Contributors				Supervisors		
	Circuit Theory	Power	Computers	Multiple Groups	Power	Computer	Multiple Groups
	N = 32	N = 57	N = 63	N = 44	N = 137	N = 88	N = 64
Organizational Interpersonal	22%	19%	24%	5%	10%	24%	19%
Organizational Documents	6	2	16	7	3	7	3
Trade Documents	25	28	16	30	30	28	25
Professional Documents	28	28	24	41	17	13	22
External-Interpersonal	19	23	21	18	40	28	31
Interpersonal Media	41%	42%	45%	23%	50%	52%	50%
Organizational Sources	28%	21%	40%	12%	13%	31%	22%

both supervisors and individual contributors. The differences among the several IEEE Groups and between the IEEE respondents in aggregate and the Electrical Engineers in the corporate surveys are reduced but not eliminated if we control for these situational factors. Tables 3-13 and 3-14 present results from laboratories and operating departments, by IEEE Group, controlling for organizational position.

These relationships are seldom strong enough to permit one to rule out chance causes. In all these cases, furthermore, the use of sources within the organization is very low, in relation to the pattern in the corporate surveys, and differences in use of alternative sources outside the organization are more marked than those between use of internal and external sources. We are thus led to conclude that, among the IEEE cases, the single most salient factor is one that is constant throughout, namely, the constellation of characteristics that leads a man to affiliate himself with a specialized technical group.

CHAPTER 4

Interpretation

WE HAVE EARLIER CHARACTERIZED THIS WORK as an attempt to analyze the *means* by which information is transferred in relation to the *purposes* of the work for which it is employed. Chapter 3 described the relationships between the utilization of information sources and particular circumstances of work. When reviewed in detail, those relationships present a complexity which, as one reader observed, "defies simple summary." The essence of our interpretation of the data, however, is that a parsimonious explanation is possible, and can be found in the data themselves. The basis of that explanation can be found in the premise that different *purposes* are associated with work done under different *circumstances*. The observed differences in utilization of information sources, then, can be explained in terms of the function of information in relation to these purposes, and not in the simple relationship to more objective personal or situational characteristics.

We shall classify the purposes of technical work according to the extent to which the work is expected to be useful along two dimensions of value: (1) contributions to the body of knowledge which defines the principles or natural laws underlying physical phenomena or manmade technology and (2) contributions to the accomplishment of particular technological ends.[1] We shall use our measures of the particular circumstances of work, i.e., task, organizational mission, professional affiliation, and personal commitment, to estimate the importance of those two aspects of the goals of work in each instance in our data.

ASPECTS OF THE GOALS OF TECHNICAL WORK

In many cases those engaged in technical work think of their work as contributing to a particular field of knowledge, such as reaction rate theory in polymer chemistry or the theory of microwave propagation. In other cases work which may be very similar in content is undertaken to contribute to the operation of real systems, such as those for production of synthetic fibers or for air traffic control. The purposes in one case have to do with the furtherance of a disciplined field of knowledge; in the other they pertain to an operational system which helps accomplish the mission of an organization.[2] Information may result from, or be used in either type of work; yet, depending on the purposes, it is likely to be disseminated through or sought within different media.

The President's Science Advisory Committee Report, *Science, Government, and Information,* characterizes these differences of purpose as constituting a "fundamental mission-discipline duality." The report identifies the relevance of that difference to the transfer of information as follows:

> The knowledge discovered by the (space program) physiologist about weightlessness is also useful to the basic physiologist interested in the kinesthetic sensory system. The basic physiologist is not likely to read, nor should he be expected to read, the space literature . . . obviously the situation is reciprocal.[3]

---

[1] These are, in principle, complementary contributions, rather than alternative or competing purposes.

[2] This is not a distinction between the work of scientists and engineers. In our sample, as in the world at large, both scientists and engineers are engaged in work of each type.

[3] President's Science Advisory Committee, 1963, p. 5.

This duality is latent in almost all technical information; it has potential value both to a developing base of systematic knowledge and to work tied to ongoing operations. Depending upon the sources from which particular information is available, it may be accessible to those engaged in one sort of work, but not to the others.

We have suggested two dimensions on which the purposes of work may differ in individual instances. One captures what might be termed the *professional* concern in technical work; the other relates to the *operations* of employing organizations. Obviously, one cannot measure the precise extent to which either purpose enters into any particular situation. But, using the measures we do have of the circumstances of work, we can make a rough assessment of whether one or the other is likely to be a dominant consideration in the conduct of the work and, by implication, in the acquisition of information. Specifically, we will classify each instance of information transfer according to the main direction of goals implicit in the following characteristics of each case:[4]

(1) The objectives of the task to which the information was applied.
(2) The mission of the organization in which the work was being performed.
(3) The character of the respondent's professional field.
(4) The respondent's personal orientation toward the content of his work.

Before reporting the results of that analysis, we shall define the basis of our scheme for classification for each of these four characteristics.

The Objectives of Tasks

The simplest scale for this characteristic would be a twofold classification of tasks as being either high or low in the extent to

---

[4] Our use of these measures assumes that implicit in each task, each organization, each professional field, and each individual's own concept of his relation to his work, is an expectation of the extent to which the work being performed will contribute to each of these dual dimensions of goals for technical work: professional and operational.

TABLE 4-1

Concepts Used to Categorize Different Tasks

		Expected Contribution to the Advancement of Knowledge	
		Low	*High*
Expected contribution to the conduct of operations	*Low*		RESEARCH Formulation or testing of scientific theories, concepts or models OR empirical investigation of physical phenomena
	High	DESIGN Combination and integration of generally available designs or components into desired products, processes, and test procedures OR refinement of existing products, processes, or test procedures	DEVELOPMENT Formulation, development, and investigation of new approaches to technical objectives

which they imply an expected contribution on either goal dimension. This yields four possible categories, as shown in Table 4-1.

For a given technical task, the expected contribution to professional knowledge would be considered high if the main goal is the development of new knowledge, and low if the goal is to solve a given problem primarily by the application of existing knowledge. We recognize that all work in science and technology draws upon an existing base of knowledge and produces, among other products, ideas, information, and other intangible outcomes. The distinction being made here, however, is intended to identify work in which there is an intent to formulate widely relevant concepts, theories, or technological principles, which will extend the existing base of knowledge.

The second dimension, indicated along the lefthand side of the table, pertains to the extent to which the work is expected to contribute to the conduct of ongoing operations. Work that is primarily concerned with the formulation of concepts and theories concerning physical phenomena would not itself directly contribute to specific operational accomplishments, whereas work intended to develop operational means of providing a desirable performance would have a relatively high expected contribution to operations.

In Table 4-1 we have filled in three of the four cells, signifying the way we classify, on these two dimensions, the specific task descriptions indicated on our questionnaire. In Chapter 3 these were categorized as "research," "development," and "design" (see Table 3-1). Research tasks have a high expected contribution to knowledge but are relatively remote from operations; design tasks fall at the opposite pole on each dimension. Relative to both of these tasks, development work occupies an interesting position. In relation to research it should yield a more direct contribution to operations, while in relation to design it has a much higher expected contribution to the state of knowledge. Development work is undertaken when there is some specific operational goal in mind, but it occurs at a phase in the R&D cycle in which first priority often is given to the synthesis, validation, and refinement of concepts and approaches.

The fourth cell in the table is left blank. The work of testing and analysis, which goes on in close association with R&D work, would seem to fit that category, since it depends primarily on the use of existing knowledge and yet is removed from the conduct of operations. We were unable to develop a satisfactory classification of the tasks that appear in this fourth cell. Since our main concern is with the research and development process, and since these tasks tend to be auxiliary to it rather than directly involved, we will proceed with only the three measures identified in the table.

Organizational Mission

Two work assignments that in an objective sense would appear to be identical can have substantially different meanings if they arise in different organizational settings. The mission of central research laboratories is substantially different from that of engineering and development departments in operating divisions, and that difference permeates the work atmosphere and shapes the goals of work in both sorts of organizations. A restatement of the principal distinction between them (as we have used it earlier) will clarify where each should be classified on our two dimensions. The central laboratories, you will recall, are units that are organizationally (and usually geographically) separate from operations; their mission is the investigation of classes of technical and scientific phenomena that bear on the missions of the corporation. In contrast, the primary goal of R&D work within operating departments is the solution of technical problems relevant to the present or future operations of that department. As this implies, we classify the mission of research laboratories as one that attributes high relevance to contributions to knowledge and lesser importance to operational considerations. The mission of the operating department falls at the opposite pole on each scale.

Professional Field

Within the scope of this study, the main distinction as to professional fields is between engineering and science. We believe that a valid distinction between these fields can be made along the lines of the two dimensions of objectives that we have been discussing, but we recognize that the distinction here is less sharp than in the preceding case. It is explicit in most ordinary definitions of science and engineering that membership in one or the other of these categories is related to the values that shape a man's work activity. That is, science, by definition, is a process of extending man's knowledge of his world, while engineering's focus of objectives is generally on operational systems, i.e., required relations between specified input and output factors.[5] We will interpret membership in a scientific professional field to imply a greater probable concern for the development of knowledge and membership in an engineering field to indicate a greater probable concern for operational results.

The Concept of Personal Orientation

Our final measure concerns an individual's personal orientation toward his work. The premise behind our use of this concept is that an individual can influence the nature of his work in a manner consistent with his own personal orientation. This occurs in his choice of places of employment, through whatever influence he can exert on task assignments, and in the interpretations he places on assigned tasks. The concept we will employ to develop specific measures is that of a "professional orientation." The general description of a person with a high professional orientation would be:

> His ability to fulfill professional goals (e.g., generate new knowledge, develop new technological approaches, and apply professional expertise to problems) is an element in the evaluation of his own self-worth. He would describe successful professional work as a source of personal satisfaction, and professional practice as a part of his identity as a person in society. In discussing aspects of life that stimulate and challenge him, he would list problems dealing with the use and extension of professional expertise. His knowledge and

---

[5] There can be a tremendous variation within each of the categories; one would expect, for example, to find a significant difference in these terms between the chemist with a Bachelor's degree working in an operating division and the Ph.D. chemist in the central research laboratory. The latter may well be "doing science," while the former, although calling himself a scientist, may be principally concerned with technology. In general, however, we feel that good use can be made of this distinction along professional lines so long as due regard is given to its limitations.

interest in the substantive and methodological problems of concern to his profession would be evident in his activities.

Finally, in discussing his aspirations and career objectives, he would include areas and skills in which he sought further development and also recognition from professional colleagues for professional contributions.

In general, the concept of professional orientation pertains to the character of an individual's identity with work and the means he uses to develop personal competence and gain recognition of his work. There is considerable literature on both aspects of this general concept. To develop an objective measure of a person's professional orientation, we need measures for two components: professional "identity" and commitment to the development of professional competence.

The set of indices used to measure professional identity has special reference to the work of Marvick, Pelz, and Barnes,[6] and to the concept of a "profession." The specific elements included in the concept of identity are (1) identification with professional goals; (2) sources of satisfaction and fulfillment; (3) focus of career aspirations; and (4) belief in the value of one's profession to society. The base for the concept of professional commitment includes the work of Gouldner, Scott, and our own prior studies.[7] A summary of the items used and a reference to the source of each item is given in Table 4-2. Taken as a whole, the construct used here differs from previous usage (1) in its inclusion of both activity and identity dimensions; (2) in the breadth of description of professional identity and the use of an explicit measure thereof; and (3) in its application to both engineers and scientists, rather than to scientists alone.

[6] D. Marvick, *Career Perspectives in a Bureaucratic Setting*, 1954; D. C. Pelz, "Some Social Factors Related to Performance in a Research Organization," 1956; and L. B. Barnes, *Organizational Systems and Engineering Groups*, 1960.

[7] See A. W. Gouldner, "Cosmopolitans and Locals: Toward an Analysis of Latent Social Roles," 1951, and C. P. McLaughlin, R. S. Rosenbloom, and F. W. Wolek, "Technology Transfer and the Flow of Technical Information in a Large Industrial Corporation," 1965. Scott's work is summarized in Chapter 3 of *Formal Organizations: A Comparative Approach*, by Peter M. Blau and W. Richard Scott. San Francisco: Chandler Publishing Co., 1962.

TABLE 4-2

Summary of the Scale of Professional Orientation

(A) Items Used for Professional Identity

Item	Source	Concept
(1) Competent professionals should devote the large majority, if not all, of their careers to their scientific or technical specialty.*	Kornhauser Barnes	Belief in the value of a profession
(2) Technical problems are generally more interesting than human or economic problems.	Barnes	Identity with professional work
(3) People with superior technical (or scientific) skills generally do not make the best associates.	Andrews	Identity with professional work
(4) The more science (or technology) one knows, the more he understands the world we all must live in.	Barnes	Belief in value of professional knowledge

SOURCES: F. M. Andrews, "An Exploration of Scientists' Motives," Analysis Memo #8, Study of Scientific Personnel, Survey Research Center at the University of Michigan, 1961. (Mimeo.) W. Kornhauser with W. O. Hagstrom, *Scientists in Industry: Conflict and Accommodation.* Berkeley: University of California Press, 1962. L. Reissman, "A Study of Role Conceptions in Bureaucracy," *Social Forces*, Vol. 27 (1949), pp. 305–310. C. R. Shepherd, "Orientation of Scientists and Engineers," *Pacific Sociological Review*, Vol. 4 (1961), pp. 79–83.

TABLE 4-2 (continued)

(B) Items Used for Professional Commitment to the Development of Competence

Item	Source	Concept
(5) A competent engineer (or scientist) is so busy that it is unrealistic to expect him to set aside more than 5% of his time for extending and developing his technical (or scientific) knowledge and skills.	Gouldner	Belief in value of professional knowledge
(6) Please think of three people each of whom is a person who stimulates your work, whose judgment you respect, and whose recognition of your accomplishment would be valuable to you. Categories for description include:	Scott Gouldner Shepherd	Professional reference group
(a) a manager or supervisor of professional activities at your organization		
(b) another professional in your own division of your organization		
(c) another professional in some other division of your organization		
(d) a person whose activities are not in science, engineering, or data processing		
(7) Extent of formal education	Gouldner	Commitment to professional expertise
(8) Extent of (a) attendance at professional meetings, (b) periodical readership, and (c) papers published	Scott Reissman	Professional activity

* Items 1 through 5 are scored on a five-point scale of agreement.

Each survey respondent was ranked as low, intermediate, or high on the scale of professional orientation according to his scores on the two component indices, "identity" and "commitment to development of competence." [8] This index of professional orientation is intended to measure one dimension of a person's career orientation as it affects his interpretation of the goals of technical work along one of the two goal dimensions defined there. Specifically, a high degree of professional orientation, as measured here, is taken to imply a high expectation of contributions to the developing state of knowledge through the conduct of technical work. It does not necessarily imply either a high or a low degree of concern for a contribution to operations. Although one would expect some sort of inverse relationship between a person's position on the dimension of professional orientation and his concern for operational goals, we conceive of this as being a separate dimension rather than an opposite pole of the same dimension. More recent studies, including work of Glaser, and Goldberg, Baker, and Rubenstein,[9] have emphasized the analysis of personal career orientation in the same sense that we use it.

The Relationship Between the Goals of Technical Work and the Sources of Technical Information

Now we can proceed to investigate the association between the means used to acquire technical information and the expectations

---

[8] Each item was scored on a five-point scale (1.0 to 5.0), and the index value was computed as a simple average of the answered items (i.e., the Likert Technique). The sample means were approximately 3.0, with standard deviations of .76 for identity and .61 for commitment to competence. Index scores were grouped in four ranges, with the boundaries placed at the approximate mean score, 3.0, and at .5 standard deviations above and below it. Individuals scoring above (or below) the mean on both indices were scored high (or low) on professional orientation, as were those with scores more than 0.5 standard deviations above or below mean on only one measure. Those with a pair of scores at opposite extremes or with both in the central range of ±.5 standard deviations were scored in the intermediate rank. Additional notes on the scale of professional orientation are given in Appendix B.

[9] B. G. Glaser, "The Local-Cosmopolitan Scientist," 1963; and L. C. Goldberg, F. Baker, and A. H. Rubenstein, "Local-Cosmopolitan: Unidimensional or Multidimensional," 1965.

about contributions to a profession and to operations inherent in the goals of the work for which the information is used. We have defined four measures that can be used to assess the extent to which the work in each instance is expected to contribute along these two dimensions of objectives (see Table 4-3). Obviously

TABLE 4-3

Score Categories on the Dimensions of Work Goals

		Expected Contribution to:	
Factor	*Category*	*Profession*	*Operations*
Task	Research	High	Low
	Development	High	High
	Design	Low	High
Locale	Central Research Laboratory	High	Low
	Operating Division	Low	High
Profession	Scientist	High	*
	Engineer	*	High
Index of	Low	Low	*
Professional	Intermediate	Unspecified	Unspecified
Orientation	High	High	*

* Probably negatively correlated with score on the opposite dimension, but not directly measured by this factor.

these are indirect measures; each is used as a very coarse method of screening into a simple ordering ("high or low"), but we believe that each is valid as a measure. That is, we believe that there is more than a chance relationship between scores on these measures and the extent to which the two underlying goal orientations are operative in each situation. The data we shall use are those reported in our corporate surveys by scientists and engineers in central laboratories and operating divisions in instances in which they described the application of information to a research, development, or design task.[10] Within these limits there were 1,083

---

[10] Specifically, this means that we have excluded respondents who were computer programmers, all respondents in the Advanced Development Division of the DATA Corporation, and instances in which the task pertained to testing, analysis, or miscellaneous functions.

instances for which we had measures on professional orientation and task. The distribution of situations in the two work settings is indicated in Table 4-4, which shows a clear difference between the

TABLE 4-4

Distribution of Instances

Within the Central Research Laboratories
$$N = 432$$
Task

Professional Orientation	Research		Development		Design		Total	
	Number	%	Number	%	Number	%	Number	%
Low	(27)	6%	(27)	6%	(24)	6%	(78)	18%
Inter-mediate	(43)	10%	(25)	6%	(18)	4%	(86)	20%
High	(133)	31%	(95)	22%	(40)	9%	(268)	62%
Total	(203)	47%	(147)	34%	(82)	19%	(432)	100%

Within the Operating Divisions
$$N = 651$$
Task

Professional Orientation	Research		Development		Design		Total	
	Number	%	Number	%	Number	%	Number	%
Low	(57)	9%	(89)	14%	(184)	28%	(330)	51%
Inter-mediate	(24)	4%	(60)	9%	(86)	13%	(171)	26%
High	(47)	7%	(54)	8%	(50)	8%	(151)	23%
Total	(128)	20%	(203)	31%	(320)	49%	(651)	100%

types of tasks and individuals found in these two quite distinct settings.[11] Within the central research laboratories, 69% of the

---

[11] The table aggregates both scientists and engineers, although there are important differences between the two types of organizations in the representation of the professions. Respondents in central research laboratories are predominantly scientists, with engineers accounting for less than one-quarter of the total. Conversely, in the operating divisions only one man in eight doing research, development, or design tasks called himself a scientist. Engineers and scientists within the same organization tend to be engaged in different types of tasks. In the central research laboratories, 40% of the engineers

respondents were doing research or development tasks and were identified as high or intermediate on the scale of professional orientation. In contrast, the large numbers in the frequency distribution are at the opposite ends of both scales in the operating division tabulation; there 64% of respondents are low or intermediate in professional orientation and are engaged in development or design tasks.

This evidence of a strong association among these four measures is not inconsistent with our expectations that (1) all four measures are related to a single set of underlying goal orientations, and (2) there is a disposition for individuals to try to influence the character of their work so that what is inherently demanded by the task, what is expected by the organization, and what is sought by the individual fall within a consistent pattern.

We can identify two sets of circumstances in which all four measures point to the same region on the underlying goal dimensions. These two turn out to be the two most common combinations. The first, designating work that could be said to have a consistently high "professional focus" occurs when a scientist with a high professional orientation seeks information for a research task performed in a central research laboratory. The other, representing the situation where information is sought for work having a consistently high expectation of contribution to ongoing operations, or an "operational focus," occurs when an engineer with a low professional orientation performs a design task within an operating division.

Figure 4.1 illustrates a striking difference between the means used to acquire information for work having these two sets of goals. When the focus is on professional contribution, the most common source of information is the professional literature. In nearly four out of five cases the information is obtained from a source outside the firm in which the work is being done. When the focus is on operational considerations, information is obtained

---

reported tasks in the design category, as opposed to 12% of the scientists, and the small group of engineers accounted for more than one-half of all the design tasks reported in that setting. In the operating divisions, 37% of the scientists were engaged in research tasks as opposed to only 15% of the engineers.

FIGURE 4.1

Modal Patterns of Circumstances

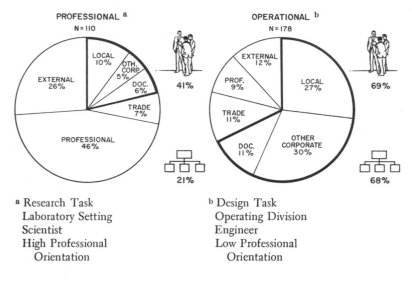

PROFESSIONAL a
N = 110

OPERATIONAL b
N = 178

a Research Task
Laboratory Setting
Scientist
High Professional
Orientation

b Design Task
Operating Division
Engineer
Low Professional
Orientation

from sources within the same firm in more than two-thirds of the cases. Although interpersonal communication is common in both cases, different sorts of informants are employed for professional, as opposed to operational purposes. Persons outside the firm represent more than two-thirds of the informants in the one case, but only one-fifth in the other.

Clustered around these two modal cases, in each setting, are a pair of similar cases; so similar that our coarse measures would not be expected to detect any difference in the character of the work. Thus, we could class as also having a predominantly professional focus work done by scientists in central laboratories in cases where a man with a high professional orientation performs a development task or a man with an intermediate professional orientation performs a research task. One would expect that the range in these categories on both the task and the professional orientation scales would be sufficiently large that they would, in fact, contain a number of situations that would fall very close to the pure "professional" category. Similarly, we can associate with the pure "operational" focus those situations in operating divisions where engineers with low professional orientation perform development work or those with an intermediate professional orientation are performing a design task.

FIGURE 4.2

SITUATIONS HAVING A PROFESSIONAL FOCUS

(Scientists in Central Laboratories)

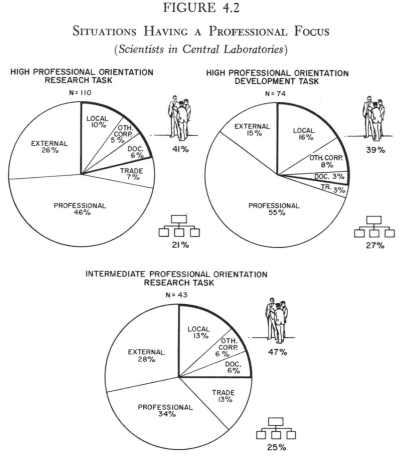

Figure 4.2 shows the distribution of means used to acquire information for the three sets of circumstances grouped within the broader classification of "professional focus" of work. Figure 4.3 presents comparable figures for the "operational focus." These

FIGURE 4.3

SITUATIONS HAVING AN OPERATIONAL FOCUS

(*Engineers in Operating Divisions*)

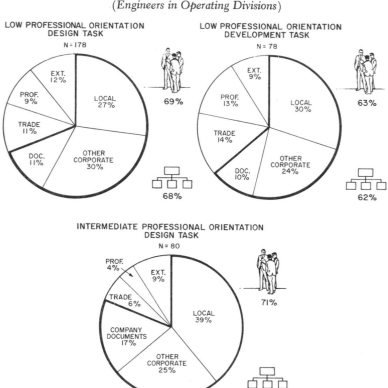

cases represent fully one-half of those under analysis. The two figures show that the pattern of information transfer is substantially the same for all three cases within each set and strikingly different between the two general situations.

The reader will note that development tasks appear in both of these supposedly different categories of circumstances. We classify development as being highly related both to the extension of knowledge and to the support of operations, and we interpret these results to mean that one or the other of these goal orientations may be given greater emphasis, depending on the circumstances.

Specifically, when a man whose own orientation is principally toward contributing to knowledge performs a development task in a central laboratory, it seems likely that the potential knowledge orientation of the task would become more salient in determining the character of the information he uses. Conversely, in operating situations, engineers (who often are less concerned with the advancement of general bodies of knowledge) may tend to place less emphasis on that aspect of a development task and considerably more on the operational consequences. In other words, we would expect that the character of the goals of work intended to "formulate, develop, or investigate" new approaches to technical objectives will vary significantly, depending upon other aspects of the work situation.

FIGURE 4.4

MIXED CASES

(*Engineers in Operating Divisions*)

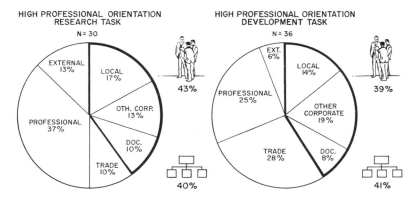

What happens when these convergent forces do not all act in the same direction? Our data provide some interesting answers, although they must be considered less conclusive than those already reported because the occurrence of such situations is considerably less frequent. Within the operating divisions there were 66 cases in which an engineer with a high professional orientation performed a research or development task. The means used to acquire information in those circumstances are depicted in Figure 4.4, which shows that the pattern of information acquisition shifts much closer to that occurring when work has a consistently "professional focus." Professional literature is used considerably more, and local, informal sources considerably less, than for other cases in the operating divisions. The incidence of use of in-house sources falls, in this "mixed" situation, about midway between the corresponding percentages for the two "pure" cases, as shown in Figure 4.1.

FIGURE 4.5

MIXED CASES

*(Men with High Professional Orientation
Employed in Central Laboratories)*

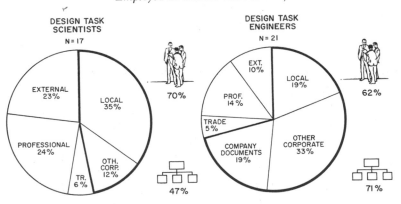

Another "mixed" case occurs in the central research laboratories when the situation involves a design task, that is, one with low expectation of contribution to the advancement of knowledge. Again, our data are based on too few cases to be conclusive, but they are consistent with the interpretation being offered here. When such tasks are done by scientists with a high professional orientation, the means of acquiring information tend to fall again in a midrange between the professional and operational pure cases. When engineers perform this sort of task, the pattern of sources shifts even further away from that associated with a professional focus in work. Both of these are depicted in Figure 4.5.

IEEE Surveys

Our surveys of members of four groups in the Institute of Electrical and Electronics Engineers (IEEE) disclosed that that population is composed primarily of senior, professionally active engineers, many of whom hold supervisory positions. The pattern of sources utilized by these respondents is nearly identical to that reported above as being associated with work having a consistent "professional focus." As reported in Chapter 3, it is sharply different from the results obtained from the most comparable sub-set of electrical engineers included in the surveys in corporations.

TABLE 4-5

IEEE Data Compared to Work with High Professional Focus

Sources	Work Having High Professional Focus* N = 110	IEEE Groups Total N = 1075	Selected IEEE** N = 246
Organizational Interpersonal	15%	17%	21%
Organizational Documents	6	6	5
Trade Documents	7	25	21
Professional Documents	46	25	32
External-Interpersonal	26	27	21
Interpersonal Media	41%	44%	42%
Organizational Sources	21%	23%	26%

* Scientists with high professional orientation, performing research tasks in central laboratories included in surveys in corporations.

** Electronics engineers (excluding Power Group) with high professional orientation, performing research, development, or design tasks in establishments of profit-making corporations.

Table 4-5 restates the results for the group of corporate responses associated with work having "high professional focus" and compares them to the pattern observed in the IEEE responses in aggregate and for a sub-set which is slightly more comparable to the corporate situations. The only notable difference between the IEEE pattern and that for work with a "professional focus" is found in the relative incidence of trade and professional documents. Across all three columns the two sources together account for roughly one-half the instances; in the case of the research scientists, however, trade documents represent an infrequent source, while for IEEE Group members they appear with almost equivalent frequency to that of the other sources outside the organization. The research scientists' work, characterized as having high "professional" relevance, is of low immediate relevance to operations (they are performing research tasks in central laboratories). The work of the IEEE members, on the other hand, must surely rank high on expected contribution to operations. They are, after all, engineers, not scientists; most of them are located in development laboratories or operating departments; and most of the tasks involved are characterized as development or design, rather than research.

This particular pattern of information transfer among IEEE Group members, predominantly directed toward sources outside their organizations but including frequent use of trade documents and interpersonal communication (which might be of greater operational relevance), can be interpreted, then, as related to the combination in their work of a high emphasis on both professional and operational goals.

As individuals, members of IEEE Groups must be ranked well above average on a scale of professional orientation. This point is supported both by our index of professional activity (see Table 3-9) and by the simple fact that affiliation with a group signifies a commitment to a specialized field which is greater than that held by most engineers. At the same time, as noted above, other aspects of their situations would lead to a high emphasis on operational goals.

Thus the typical IEEE Group member may exemplify that situation in which the goals of work place emphasis both on contribution to professional knowledge and relevance to ongoing operations. This interpretation, which is speculative rather than being testable within the framework of our data, would account for the two main features of the IEEE results. The very low incidence of use of sources within the organization, in contrast to the higher proportions typical of engineering groups in the corporate surveys, can be interpreted as a concomitant of work with high professional focus. Yet, in relation to the pattern of sources associated with professionally focused work performed by scientists in central laboratories (where operational considerations are diminished), the IEEE results show a high incidence of use of trade magazines, technical reports, commercial literature, and similar documents. This difference could be a reflection of the coordinate concern for relevance to operations, which may not be met through the alternative use of the professional literature.

SUMMARY

This chapter has considered the question, "Under what circumstances are the various means for information transfer used to acquire information in support of research and development tasks in industrial organizations?" The principal distinction as to means, based on the concerns of managers and policy makers, seemed to lie between local and informal sources on the one hand, and formal sources outside the firm (principally the published literature) on the other. Our data show that when the circumstances of work are such as to create a high expectation of contribution to a developing body of knowledge, i.e., a professional focus, information transfer will be predominantly through the formal media. When the focus of work is principally upon intended contributions to ongoing operations, the transfer of information will occur primarily through informal media and will involve predominantly local sources. This finding is based only on cases in which engineers and scientists were doing research, development, or design tasks in industrial laboratories or engineering departments. The most persuasive findings arise in circumstances where one focus of goals is considerably more salient than the other. The expected change could be found, however, in the less frequent cases where the indicators we used were mixed. There was no substantial number of cases in the data from the corporate surveys in which both operational and professional goals could be expected to be highly salient.

A speculative interpretation of the results of the surveys of IEEE Groups, however, would explain that distinctive pattern of sources for information transfer as a concomitant of a simultaneous concern for both professional and operational contributions.

CHAPTER 5

Conclusion

IN SCIENCE AND TECHNOLOGY, the expectations of professional practice imply that each contributor should build on the work of predecessors and colleagues and that each branch, or discipline, interact with other relevant branches. Technical progress, in other words, is an accumulation of advances in the frontiers of knowledge or the "state of the art," built on the most current information about relevant knowledge and practice. An ideal information system, therefore, would ensure that relevant information, whatever its origins, is accessible to meet whatever needs arise in technical work.

Practical considerations imply that the goal of making information "accessible to all" cannot be realized fully. Nor does it seem realistic to imagine that work in any field or organization can be served by *an* information system. For example, our data on the circumstances leading up to the acquisition of information show that a great deal of technical information is transferred in the absence of a fixed and focused plan on the part of its receiver. Half of the incidents reported to us involved the discovery of specifically useful information at a time when either more general competence-building knowledge was being sought or when it was pointed out on the initiative of another individual. Not only must we ask how people find information, but also how information finds them.

More important to managers is the clear evidence of the importance of what we might call the "culture of work" in science and technology. The professional communities of science and engineering are characterized by traditions, institutions, and social relationships which define many of the channels for the transmission of knowledge and contribute to the steady enlargement of its domain. The values that lead one man to seek to keep up with his field, or the social bonds that lead another to call his attention to certain information, are as much a part of the information transfer system as are conferences, journals, and documentation systems. The manager who would understand and influence the process of information transfer should consider its elements as functional components of a single system which includes these social and institutional elements.[1]

The main finding of this study is that the propensity to use alternative types of sources for technical information is related to the purposes which will give meaning to the use of that information. We found that work which has a professional focus draws heavily on sources of information external to the user's organization (i.e., on the literature and professionals employed in other organizations). Conversely, work which has an operational focus seldom draws on external sources, relying heavily on information which is available within the employing organization.

Among the many diverse forms that work with a "professional" focus assumes, the unifying element is an intent to produce a contribution to knowledge. The product of such work must, by

---

[1] The importance of such a point of view has been emphasized by Herbert Menzel in "Scientific Communication: Five Sociological Themes," *American Psychologist*, Vol. 21 (1966), p. 1000 as follows:

It is necessary to look upon any one arrangement, institution, facility, or policy for scientific and professional communication as a component of the total system of scientific communication for a profession, a system which includes all the provisions, all the publications, all the facilities, all the occasions and arrangements, and all the customs in the discipline that determine how scientific messages are transmitted.

definition, be original, understandable, and of value to others. The "others" in question comprise the group of peers in that professional field. Information is useful for this work if it helps to *define* an existing frontier of knowledge, to *verify* the interest of others in the area, or to *clarify* available results in relation to existing theory and concepts. The various items of information sought in such circumstances have a single point in common: their relevance is defined in relation to a recognized body of knowledge. Knowledge knows no organizational bounds; the relevant body of knowledge is found in a literature which is open to all organizations and a set of colleagues whose work is conducted in many organizations.

Effective performance in the development of operational technology also depends upon knowledge, i.e., on systematic understanding of natural phenomena and manmade devices. However, in this case the knowledge does not become useful until it is translated into terms relevant to a *need* for the technology and a *capability* to produce it. Information is valued for operational purposes if it is expressed in the context of a technological system. In that way knowledge about an isolated phenomenon can be related to client needs or an established production capability, as well as to knowledge of the performance of other component technologies which will comprise an overall system.[2]

In brief, for professional purposes, the primary focus is knowledge, regardless of its scope and influence, while for operational purposes the primary focus is the context in which knowledge may be applied. Thus, when those engaged in professional work told us why they valued certain information sources, they would commonly emphasize the importance of simplicity, precision, and analytical or empirical rigor. Conversely, those engaged in operational work typically emphasize the value of communication with others who understand and are experienced in the same real con-

---

[2] As already noted in Chapter 1 (see the discussion of Scott's comments on pages 8 and 9), some people take a different view about the relative use of the literature in operations-oriented work. That is, some assume that the pattern of reliance on local sources indicates the technologists are at "fault." The data obtained in our study cannot resolve this difference concerning the apparent provincialism of such groups. To understand this question, one would have to examine both the patterns of information use and the nature of operations-oriented work, a task which lies outside the scope of this effort.

text of work. In each case one finds that a certain unity of purpose is expressed in the task and its setting, in the professional affiliation and career orientation of those performing it, and in the character of the information sources most commonly employed.

The distinction between professional and operational purposes, viewed against the common characteristics of technical organizations, can help to explain the observed differences in the utilization of information sources for the two kinds of work. In the technical organizations of most industrial firms, a man seeking to contribute to the extension of a field of knowledge will necessarily find that few of his colleagues in the organization orient their work toward the same particular field. There is a myriad of ways of dividing man's quest for knowledge and no special license which limits its advance to any one group. As a result, only a small fraction of the work in any basic field is performed within a single organization. The unifying element between the work of individuals and groups in different organizations is their common way of defining their field and the factors which are important to its extension. The unity is abstract and lies in the logical structure of the discipline being developed. It is not surprising to find that, under these circumstances, information is more frequently obtained from the open literature and distant colleagues than from local sources.

The predominant use of local information sources in operational circumstances is often attributed to the higher "costs" of acquiring information by other means. That is, it can be explained as a manifestation of the simple economics of turning around (or picking up the phone) and "asking the next guy." The most notable point about operating organizations, however, is that they ensure that usually there is a "next guy" to talk to. A design engineer, looking for others sharing his concern for the application of a technical expertise to a specific operational need, will find that the work of many of his immediate colleagues is directly relevant. In fact, such establishments often express pride in the belief that they have become specialized, experienced, and professional problem-solvers for certain kinds of operational needs. The same characteristic of this work which imparts value to information of certain kinds also makes it worthwhile to hire and organize professionals to create effective working teams. That is, the unifying element in such work is provided by the mission of the organization, which not only relates task to task and person to person, but which, when fulfilled, also justifies the development and support of a team of experts.

IMPLICATIONS FOR MANAGEMENT

So far we have emphasized the differences between professional and operational purposes in technical work and the significant effect which a predominant focus on either purpose has on the utilization of information sources. The manager in a mission-oriented organization, such as a large corporation, often is concerned with a system that includes both kinds of purposes. From the viewpoint of the system as a whole, his aim must be to contribute to the unity of both kinds of work. This is especially relevant in the mission-oriented organization that sponsors scientific research or undertakes technological developments which push into areas of newly discovered knowledge. In the former case, the choices taken in professionally-oriented scientific work are expected to be influenced by the operational mission of the organization; in the latter the most advanced knowledge is likely to be needed in solving operational problems. The manager, then, must be concerned with measures which complement and interrelate the regular patterns of information flow. To do so he must build and maintain within his organization an effective interface between information systems concerned with advancing bodies of knowledge and those concerned with operational systems.

Where there is no effective interface, the situation may appear as a caricature of the process outlined here. Professionally-oriented work will build on information obtained from the literature and from direct personal contacts with professional colleagues. The results of such work feed back into those same information channels, thus closing the system on itself. Operationally-directed work, meanwhile, will draw upon experience-based information largely derived from sources within the same firm. Again, the system is closed, as the results of such work feed into the same local channels, but fail to cross over into the professionally-oriented system.

The problem of building a practical means for information transfer between discipline and operations-oriented work is a very demanding and challenging task, as suggested by the following statement from a governmental task force:

> One of the features that characterizes the growth of science and technology is the increasing complexity of interrelations among its elements. . . . For example, research on detection of infrared radiation is related to the field of optics since it deals with optical radiation; to the field of solid state physics by virtue of its contribution to understanding of the solid state; to chemistry by virtue of its potential contribution to spectrometric analysis of structure and composition; to instrumentation as a contribution to the transducer art; to communication by means of infrared radiation; to missile guidance; to aircraft safety; to control of space vehicles; to military surveillance; to criminal investigation; to identification of works of art; and to innumerable other areas and in ways not yet conceived.[3]

In our view the key to establishing an interface is contained in the idea of a unifying purpose which we used earlier to explain the differences between the two patterns described by our data. Underlying the pattern of information transfer between different people in either an operational area or a discipline, there is a common purpose which lends unity to tasks which might otherwise appear to be separate and unique. The question which concerns us in this section is that of defining the still higher purposes which could structure and unify a system which included both of these patterns. In our view that kind of purpose can only be defined by management; it is not intrinsic in the tasks of the professionals involved.

Organizing Task Assignments

Within a large organization the location of work groups and the definition of organizational relationships between them has an obvious effect on the patterns of information transfer. By deliberate

---

[3] J. H. Crawford, et al., *Scientific and Technical Communications in Government,* Task Force Report to the President's Special Assistant for Science and Technology, April 1962 (AD-299-545).

assignment of certain technical tasks to particular individuals and groups, management can create linkages between the professional and operational work within the organization. One such technique involves the creation of work in which both professional and operational purposes are combined, as, for example, would occur when men with high professional orientations undertake work with high professional contribution but perform it within an organization primarily concerned with operational responsibilities. The Ph.D.s doing advanced development work in an operating division represent one such assignment. Many operating departments also include groups which perform analytical studies for and consult with project groups. These special analysis groups (e.g., a "network synthesis" group for electrical systems, a "stress and vibrations" group for mechanical systems, or a "corrosion analysis" group for chemical-process systems) not only allow professionally-oriented individuals to pursue advanced work, but reward such work as it relates to the ongoing context of mission-oriented projects.

When management allows development work to be clearly based in a discipline while providing an equally clear relation to an organizational mission, the resultant structure should facilitate the transfer of knowledge. In the Bell Laboratories, for example, the development organization is located close to the research department, while being tied organizationally to design activities (which are, in turn, located near manufacturing operations). The simple pattern of formal and informal ties and barriers thus created seems to serve that organization well.[4]

Development sections might also be attached to research laboratories. While this is possible, however, it must be noted that it is the rare individual who has a high concern for both professional and operational matters. Furthermore, experience has shown that management must take great care about communicating the role and scope of such groups in a laboratory. In the absence of any clear function for these groups, they can become the focus for disruptive, organizational disputes.[5] It may also be difficult to

---

[4] J. Morton, "From Research to Technology," *International Science and Technology*, May 1964, pp. 82–92.

[5] A case history which illustrates the kind of dysfunctional disputes which may ensue is that written by C. R. Shepherd and P. Brown, "The Impact of

organize professionally-oriented and competent individuals in an operating division. These difficulties may force management to adopt a different strategy.

One way of alleviating the difficulties of finding people who can adopt a dual orientation arises naturally when a research project yields applicable results. Many discipline-oriented people are interested in following a project ("their baby") out of research, one or two steps closer to application in the operational world. Similarly, mission-oriented people may find consistencies in their work across projects, or they may develop generalizations about technological phenomena and be interested in taking these one or two steps along the path to rigorous empirical study or systematic analysis.[6] Another approach is to create a separate organizational unit which is charged formally with advanced development responsibilities. In our own study, the one organization primarily devoted to such tasks, the Advanced Development Laboratory in DATA Corporation, showed the greatest incidence of interpersonal communication with persons employed in other units of the same firm and also an above-average use of literature sources. Although that result might be explained in other ways, it is consistent with the performance of a "linking" function by the laboratory.

Organizing Document Systems

This study is based upon an exploration of the interface between task and information, rather than between user and information service. Yet we would be remiss if our discussion of management implications did not refer to the technical literature. Documents are more than a record which may be catalogued, indexed, abstracted, and stored. The heart of the document, its technical content, is an expression of the organization of one area of advancing knowledge. As such our concern with the user and his

---

Altered Objectives: Factionalism and Organizational Change in a Research Laboratory," 1956.

[6] Although it is often not recognized, a scientific discipline may spring from the accumulated work of engineers and operationally-oriented people. The discipline of Strength of Materials is one such example. More recently the disciplines of Control Theory and (Petroleum) Reservoir Dynamics have arisen from the work of engineers and petroleum technologists.

work is usefully turned to a discussion of the user and the structure of available, documented knowledge.

At the beginning of this chapter we emphasized one of the main points of our findings: information is frequently found not as the result of a direct search related to a specific task, but as part of a more general quest for knowledge which will enhance the user's background or keep him up to date on professional developments. In this "general competence" mode of acquiring information, the literature plays an especially significant role.[7] Thus there is an even stronger reason for investigating and acting on the identity which literature sources have and the structure which relates important sources to each other.

Our survey indicates that the membership of IEEE professional groups constitutes a group of individuals who are highly professionally-oriented even within otherwise operations-oriented establishments. However, their concern with the literature as a vehicle for the transfer of information is expressed in their reliance upon trade magazines and other trade documents, not professional journals and books. This finding suggests the possibility that some literature sources can perform the function of transmitting new professional progress to audiences with special operational interests.

The function of such "intermediary" documents is probably not confined to that of a distribution medium. If that were their only function, the oft-heard complaints about "meaningless duplication" in the literature would be justified and worth acting on. But these sources probably do more than just repeat information that is otherwise available in the same form in professional sources. We noted above that one of the basic attributes of operationally useful information is its clear relevance to the context of technical developments. Thus, intermediary journals can make a unique contribution; the authors of papers which appear in them may *add* new knowledge in the form of context-relevant statements. Such statements would relate knowledge about natural phenomena or

---

[7] Our own data (see Figure 3.5) show that for scientists, 79% of the competence-oriented incidents involved the use of literature sources (64% were professional and 15% trade sources); for engineers, 53% of the incidents involved the literature (26% professional and 27% trade sources).

analytical theory to technological needs, or to capabilities for producing devices, or to already applied principles and bodies of empirically derived performance data. However, such statements need not deal with natural phenomena or theory at all; technical performance trends may be related to each other; in other words, the union of professional knowledge and operations need not spring solely from science.

In his study of literature citations, Derek Price[8] has commented on the "historical independence" of science and technology as follows:

> Science and technology each have their own separate cumulating structures. . . . It is probably that research-front technology is strongly related only to that part of scientific knowledge that has been packed down as part of ambient learning and education, not to research-front science.

It seems likely that "packed-down" knowledge is important in the application of knowledge to operational use. In addition, however, we would emphasize the positive aspect of the technologist's literature: the addition of new knowledge which places basic learning closer to an operational context. The bridge between discipline and operation-oriented progress undoubtedly does lag the advancing front in both areas, but this is because this bridge is also based on new knowledge. The more one has the knowledge to relate performance capabilities to general principles or to relate research findings to the complex particulars of an operational system, the stronger will be the tie between the two areas.

The logic underlying the organization of the open literature is not the logic of a mission-oriented program, but that of a body of accumulating knowledge. Each source fits into a tradition of inquiry and concern which has its own list of central issues, base of empirical experience, and context for generalization. The problems of having too much to read or of needlessly duplicating someone else's work are not problems of the size of the literature, but problems of the organization of the literature. The key issue, in our

---

[8] D. J. de S. Price, "Is Technology Historically Independent of Science?" *Technology and Culture*, Vol. 6 (1965), p. 568.

judgment, is the user's ability to judge the relevance of a source to needs, not that of making one journal or source relevant to all.

The problem of judging the relevance of a source centers on the identity which users attach to available documents. To some extent, this identity is in the hands of the professional groups and publishers which sponsor the literature. An example of one effort at literature organization is the set of journals maintained by the American Psychological Association. Not only are these journals differentiated according to subject, but also according to type of contribution (e.g., review of the literature, empirical study, theoretical analysis, etc.). In the final analysis, however, the identity attached to each source and each item of literature will be a function of the particular objectives which characterize a given organization. Groups which have the responsibility for storage and retrieval can express their appreciation of this fact by using different cataloguing and indexing systems for different sources rather than a universal (and probably discipline-based) system for all. The management of mission-oriented organizations may also play a role not only by supporting the formation and operation of associations which sponsor the development of relevant knowledge, but by taking the responsibility for sponsoring important document sources themselves.

The model we must keep in mind is not a two-step model relating discipline to operation, but a network composed of many nodes and links.[9] In some fields (e.g., the early stages of transistor research) management may find that little context-oriented knowledge is needed to link the most abstract discipline to commercial operations. In such cases only one intermediate source may be needed.[10] In other cases (e.g., machine design) much in the way of context-oriented information will be needed and several sources

---

[9] This idea of a network also indicates the value of sources whose focus is on interoperations or interdisciplinary transfer as well as on discipline to operations transfer.

[10] In an analysis of literature in physics and electrical engineering, M. M. Kessler, "Technical Information Flow Patterns," pp. 249–250 (Mimeo, undated, AD-261–303), found that for the subject of transistor research, the usual pattern of industrial and university contributions to the highly abstract *Physical Review* was completely reversed, so that industrial contributions predominated.

may be used to fill the gap. In some cases the gap will be a unique problem for a particular company, and in those cases that company could use its own internal documents and literature to fill it. The key point to keep in mind when considering such "internal" documentation is the role of the source. The idea is to produce literature which has an identity and fills a need, not a mere imitation of already existing sources.

In some companies there is a failure to provide easy access to many professional sources on grounds that they only contain "worthless efforts at hairsplitting." In some professional groups there is an outcry against the publication of "trash and duplication." The point that we stress is that "trash" and "worthless" are epithets applied to work which has little relevance to its reader, not necessarily little relevance to some important part of the scientific-technical community. Once again the problem lies not so much in publication as in the clarity of each source's relevance to its users.[11]

Organizing People

Our findings have shown the important part that people play in the information transfer process both by providing information directly and by referring users to other sources. The organization within which technical work is done is, of itself, a principal mechanism for the storage and transfer of information. Managers affect that system in numerous ways. The relevance of actions affecting the availability of libraries or other information services is obvious; other management actions, concerning job definitions, staffing, organizational structure, and the like may often seem unrelated but can have equally significant effects on the transfer of technical information.

To illustrate this, we must move beyond the hard evidence of survey data to discuss facets of information flow which were prominent in our discussions with respondents and other scientists and engineers. Within a large technical organization, and between organizations, informal relationships interact with the formal struc-

---

[11] The point made in this section may also be applied to often-heard calls for closer refereeing and even censorship of the literature, with the intent of eliminating "wasteful trash and duplication."

ture to yield a new matrix of interconnections between people and tasks. Many of those connections are realized through the social fabric of life within an organization, that is, as a result of contact between friends and acquaintances. Others are realized as a result of professional interests which lie outside the framework of work assignments. Scientists and engineers do not "turn on" their special approaches and skills only when they are working. People who enjoy and are committed to their identity as practitioners in science or technology look ahead and around and find many subjects which fascinate them. The matrix of these informal interests, many of which, though not formally a part of work, nevertheless do derive from and are related to work, provides the structure for important networks for information transfer.

It is a contradiction in terms for managers to control the informal and social fabric of their organizations. Yet, management may have an important influence, either positive or negative. Through established procedures and policies, managers can affect the ease with which people can make and maintain informal contacts. Policies relating to travel, guest visits, and telephone calls are an important aspect of this picture.[12] A particularly significant factor may be the way management recognizes the legitimate, but not directly work-related interests of its personnel. Encouragement and support for these personal interests may be shown in opportunities for training, the availability of literature, and freedom for travel to important sources of stimulation and knowledge. Insofar as such efforts recognize not only the technical interests of people, but also interests in the context of corporate operations, an important aspect of the bridge between the worlds of discipline knowledge and operational practice may be formed. It should be noted that this bridge is particularly useful not only for the direct transfer of information, but also for those other important circum-

---

[12] Management may also influence this aspect of organization through their investments in facilities for coffee-breaks, luncheons, and meetings. One interesting example of such a concern with lunchroom facilities was a company which made a careful study of the kinds of paper which could be used for napkins. The problem is to find a paper which serves the function of a napkin and yet has sufficient body to serve as a sketch pad for notes and diagrams.

stances, the transfer which results during competence-building reviews and the transfer which results because a source voluntarily passes information on.

The media for information transfer which emerge from the structure of the organization can be valid and effective mechanisms for meeting the needs of professionals. Although this sort of communication may appear to be haphazard and perhaps inefficient, it seems more useful to recognize that it may be the natural functioning of a highly sophisticated and well-structured social system. Provincialism is not a necessary result of reliance on local and informal sources. Whether or not it is the result will depend, in our opinion, upon the clarity and depth to which executives understand the many interrelated transfer systems which they manage and influence.

Conclusion

Both research laboratories and operating organizations can share in the common task of a mission-oriented technical organization: learning relevant new knowledge and applying it. The problem common to both activities is one of maintaining the linkage between awareness of the possibilities of new knowledge on the one hand, and awareness of the needs of operations and the marketplace on the other. In industrial research the major problem is avoiding irrelevance; in design the problem is one of avoiding obsolescence. In central laboratories, which tend to be well coupled to the dynamic world of science and technology, the problem is made manifest as one of maintaining contact with the real needs of the organization. In operating units, which tend to be well coupled to the real world of needs and capabilities, empirical success in making something work can seem more important than understanding why it works. The problem in that environment is one of maintaining contact with knowledge of the best techniques.

These two problems, maintaining the relevance of research work and avoiding obsolescence in engineering work, are aspects of a central question for R&D management: how to maintain fruitful linkages between the sources of knowledge and the needs for knowledge in operations. In this chapter we have tried to outline the context in which positive efforts to support the necessary transfer process may proceed. We believe that definition of what is needed in a given situation is the proper concern of professionals in information science and management and should be guided in each case by studies of the particulars of the situation. We have attempted to provide a rationale with which management may understand the need for and the work of such professionals.

The context for information transfer policies which we have presented applies to management at all levels. We have noted the important role of the formal organization of a company which relates division to division as well as man to man. Furthermore, we have emphasized the influence of all levels of management on the opportunities which people have to follow their professional interests and to share them with others who might both learn from and contribute to them. In addressing the network of literature sources, we have encouraged the active role of professional societies and directors of publication in searching out and filling needs for media which can relate discipline to discipline or professional concern to operational reality. When such efforts are conducted with the support of large corporations, they may be a powerful contributor to the unity of technical work.

Finally, our analysis has implications for the individual engineer and scientist on whom the responsibility for appropriate information transfer rightfully rests. Through his efforts to understand the program in which his tasks are set as well as his efforts to develop identifiable expertise and interests, the individual provides the core of the information networks which relate him to others. Through his efforts to enrich his background and appreciate the contributions of people throughout the spectrum of pure theory to market use, the individual gives himself the flexibility and mobility he needs to make the transfer process work.

Appendixes

Bibliography

Appendix A

Survey Questionnaires

Each questionnaire consisted of four principal sections:

(1) A cover letter.
(2) A page of instructions.
(3) Thirteen questions pertaining to the instance being reported.
(4) A set of background questions pertaining to the individual respondent.

IEEE Surveys

The two IEEE questionnaires were identical except for the cover letter and three screening questions added to the second page of the second mailing. This appendix contains the cover letter from the first questionnaire and the second questionnaire in full.

Corporate Surveys

Each questionnaire used in the corporate surveys had on its cover a memorandum on the corporate letterhead signed by the engineering manager, research manager, or general manager of the establishment concerned. A typical cover letter is illustrated here following the IEEE questionnaire. The instructions were the same as for the IEEE, except that in the third paragraph the phrase "someone in your group" referred instead in some versions to "section," "unit," or "department," depending on the terminology in the establishment concerned.

Questions 1–13 on the corporate questionnaires were identical to those appearing on the IEEE questionnaire, with these three exceptions:

(1) Answers (a) and (b) to question 6 (Source) were phrased to fit the corporate organization structure, as shown by the following version used in units of DELTA Corporation:
  (a) An employee in my own Engineering Department; his name is _____.
  (b) Another DELTA employee in the _____ department of the _____ division.
(2) Answers (a) and (c) to Question 7 (Lead) were phrased to parallel (a) and (b) of Question 6.
(3) The task description in Question 10 differed slightly in some versions to fit varying technologies and types of work.

The background questions used in the corporate surveys are illustrated here by Part II of the questionnaire used in the DATA Corporation.

THE INSTITUTE OF ELECTRICAL AND ELECTRONICS ENGINEER

INCORPORATED

345 EAST 47th STREET • NEW YORK, N.Y. 10017
PL 2-6800

Dear IEEE Member:

The IEEE is working with other scientific and technical organizations with the goal of providing better information service. You may be aware of our efforts with the Engineers Joint Council to develop a coordinated information retrieval system. Several of the IEEE Group Transactions have started publishing index cards. Another attack on this problem is being made from the viewpoint of the management of scientific and engineering activities. What are the sources of new ideas and concepts, relevant data and specifications? What are typical information flow patterns? What uses are made of information resources within an organization? It is obvious that better answers to these questions might show how IEEE could significantly improve its service to individual members and the profession.

With the above in mind, I am writing to you, asking that you help through anonymous participation in a study being conducted by the Harvard Business School under a grant from the National Science Foundation. The questionnaire on which this study is based is being sent to a small sample of IEEE members, selected on the basis of certain Group memberships and random geographical distribution. Authorization for the distribution of the enclosed questionnaire has been obtained from the Chairmen of the Groups involved.

The questionnaire has been designed to obtain data on how engineers and scientists <u>employed in organizations</u> obtain information which is useful in the conduct of their work. If the primary nature of your own work does not involve scientific and technical activities, or if you are not employed in an organization (i.e., if you work as an individual consultant rather than in a university, company, government agency, laboratory, etc.), please check the box in the lower right-hand corner of this page and <u>return</u> the unanswered questionnaire in the enclosed envelope. If you <u>are</u> actively engaged as a scientist or engineer employed by an organization, I urge you to answer the questionnaire.

During the 15 to 20 minutes required to fulfill this request, you should treat this piece of research with the same respect you would accord to research or design work in your own field. This questionnaire has been worded to safeguard both your organization's proprietary information and your individual interests. Your answers will be analyzed and reported only in aggregate statistical terms. Please complete the questionnaire at one sitting and return it using the enclosed envelope, within the next <u>ten</u> days.

When the study is finished, the results will be sent to the IEEE in order that we may immediately and directly act on any of the findings that would improve our service. Thus, I am asking that you personally help us and thank you, in advance, for your important contribution to this effort.

Cordially,

D. G. Fink
General Manager

My activities are not primarily in science or engineering or I am self-employed. ☐

HE INSTITUTE OF ELECTRICAL AND ELECTRONICS ENGINEERS

INCORPORATED

345 EAST 47th STREET • NEW YORK, N.Y. 10017
PL 2-6800

September 15, 1965

Dear IEEE Member:

Several months ago we agreed to help the Harvard Business School with a study of technical information flow; the study is supported financially by a grant from the National Science Foundation. The IEEE help consisted of voluntary, questionnaire responses by a sample of IEEE Group members, drawn at random but presumably matching the professional interests of an "industry population" the Harvard Business School obtained through other arrangements. I asked some 2400 of our members to help, and over 600 completed and returned the questionnaire.

The results obtained from an analysis of these returns were unexpected. The 600 IEEE respondents have some information utilization characteristics that are statistically different from the usual pattern. The question immediately arises whether the 600 respondents were typical of the initial 2400 population, and hence of IEEE Group members. Fortunately, when we compiled the original sample of 2400 names, we prepared a duplicate set of envelopes in order that we could later communicate with the same members if a need arose. This has now happened. The Harvard Business School has suggested, and I have agreed, that we use the reserve set of envelopes for a study to determine the internal homogeneity of the 2400-population. If the profile of the 600 respondents proves to be valid for the entire 2400 population, we will have a result that is significant to the Harvard Business School study and, I assure you, of great interest to IEEE.

I dare not say more, for fear of prejudicing the homogeneity study that is being attempted. I earnestly beseech each of you to respond to the enclosed questionnaire, knowing that I am asking 600 of you to act twice. A summary of the study will be prepared for publication in your respective Transactions and Newsletters.

The enclosed questionnaire is designed to safeguard individual and organizational interests. The survey is anonymous and responses will be analyzed and reported only in aggregate terms. The questionnaire may require only ten minutes of your time and I hope you will accord it the respect you give to professional work in your field. Please return the questionnaire, in the enclosed envelope, by **October 15.**

Your sympathetic understanding of our situation will be greatly appreciated and I want to thank you in advance for your assistance and continued contribution to this effort.

Sincerely yours,

Donald G. Fink
General Manager IEEE

Please answer these questions before proceeding with the rest of the questionnaire; in each case check the ONE most c[...] plicable box.

1. I am:

a. self employed .... ☐ (55-1)
b. employed in an organization (i.e., agency, company, educational institution, etc.) .... ☐ (55-2)
c. other (please describe) _____ ☐ (55-3)

2. My work primarily consists of:

a. direct involvement in scientific and engineering tasks .... ☐ (56-1)
b. supervision or management of scientific or technical groups .... ☐ (56-2)
c. teaching science or engineering .... ☐ (56-3)
d. activities other than those found in science or engineering .... ☐ (56-4)
e. other (please describe) _____ ☐ (56-5)

3. Please check one of the following:

a. I did return a questionnaire from the previous mailing .... ☐ (57-1)
b. I did **not** return a questionnaire from the previous mailing .... ☐ (57-2)
c. to the best of my knowledge I did not receive an earlier version of this questionnaire .... ☐ (57-3)

Regardless of your answers to the above questions, please read the instructions on the opposite page and then complete the questionnaire.

SURVEY OF TECHNICAL INFORMATION FLOW

INSTRUCTIONS

The questions on the following pages ask you to describe the most recent instance in which a technical idea or item of information, which you obtained from a source outside your immediate circle of colleagues, proved to be useful in your work.

Our objective in asking you these questions is to obtain data which will describe the sources, channels, and references to information which are used in the conduct of technical work. While such data may be of use in the design of information systems, our primary goal is to contribute to an understanding of how technical work gets done. This study is intended to obtain descriptive data on how technical information enters your work and how it bears on the technical decisions you regularly make.

The reference to **an item of information useful in your work** is meant to include the very wide variety of information (i.e., concepts, data, ideas, specifications, etc.) which may have been useful in your work. However, this survey is concerned only with the flow of technical and scientific information which helped you understand how to accomplish an objective. It is not concerned with the way you obtained job assignments, work schedules, or budgets.

The specification of **"source outside your immediate circle of colleagues"** is meant to limit your answer to instances in which the substance of the information came from someone who was not a member of the same basic formal unit of organization of which you are a member. By this we mean to exclude the small group of colleagues reporting directly to the same supervisor, or, if you are in a supervisory position, the group whose work you supervise directly. If the substance of the information came from an outside source to which you had been referred by a person within your immediate group of colleagues, you should include the instance.

In this same vein, you should exclude items of information which you obtained from documents which are routinely used in your everyday work (i.e., the handbooks, catalogs, and specification folders which you habitually use). Information obtained from such documents should be included when the instance of their use fell outside the context of established routine.

"Most recent" should be interpreted strictly. Our purpose is to obtain a cross-section of the information received during a limited time span. If you are debating whether to report a minor item you learned of this morning or to refer instead to some earthshaking discovery made yesterday, please describe this morning's example.

Thank you for your cooperation.

Please think of the most recent instance in which an item of information, which you received from a source other th
someone in your immediate circle of colleagues, proved to be useful in your work. Answer the following questions w
reference to that instance. In answering the multiple-choice questions, please check only the one box most applicable

1. Today's date is _____

   The date on which this instance occurred was _____

2. Before receiving this information I:

   a. had recognized a need for such information .... ☐ (7
   b. had not recognized a need for such information .... ☐ (7

If you checked category a above, please answer question 3 but do not answer question 4.

If you checked category b above, please skip question 3 and do answer question 4.

3. I obtained this information because:

   a. I specifically searched for the information .... ☐ (8
   b. someone gave me this information, a lead to it, or the material containing the information on the basis of
      having been previously told of my interests in such information .... ☐ (8
   c. someone gave me the information, a lead to it, or the material containing the information on a voluntary
      basis .... ☐ (8
   d. I ran across it or a lead to it while searching specifically for some other item of information .... ☐ (8
   e. I found it while reviewing current technical or scientific news .... ☐ (8
   f. I located it while making a search of the literature or survey of information sources in a technical or
      scientific area .... ☐ (8
   g. after recognizing the need, I went directly to a document or person from which I expected to find it .... ☐ (8
   h. after recognizing the need, I went directly to a document or person from which I expected to find a lead
      to the information .... ☐ (8
   i. of the following (please describe) _____
      _____ ☐ (8

GO TO QUESTION 5

4. I obtained this information or an oral or written lead to it:

   a. while searching specifically for some other item of information .... ☐ (9
   b. while reviewing current technical or scientific news .... ☐ (9
   c. while brushing up on a technical or scientific field .... ☐ (9
   d. while trying to learn a new field .... ☐ (9
   e. because someone voluntarily pointed it out to me .... ☐ (9
   f. while making a search of the literature or survey of information sources in a technical or scientific area ☐ (9
   g. in a manner not listed above (please describe) _____
      _____ ☐ (9

CONTINUE WITH QUESTION 5

5. I first encountered the substance (actual data, theory, concepts, or methods) of this information

   ORALLY IN: **IN WRITTEN FORM IN:**

   a. an informal face-to-face conversation .. ☐ (10-1)   d. a published book .... ☐ (10
                                                           e. a published technical or scientific journal
   b. a telephone call .... ☐ (10-2)   or conference preprint .... ☐ (10
                                                           f. a published trade magazine .... ☐ (10
   c. a prearranged meeting, conference or   g. a report of my own organization .... ☐ (10
      seminar .... ☐ (10-3)   h. supplier catalogs or technical trade literature ☐ (10
                                                           i. a technical report issued by another organ-
                                                              ization (e.g., company, university, et al.) ... ☐ (10
                                                           j. a document not listed above (please describe)
                                                              _____ ☐ (10

 y informant or the principal author(s) of the document which contained the substance of the information,
as (were):

- an engineer or scientist employed in my own organization . . . . □ (11-1)
- a non-technical employee of my own organization . . . . □ (11-2)
- an engineer or scientist from another company (other than a sales representative) . . . . □ (11-3)
- a sales representative of another company . . . . □ (11-4)
- another employee of another company . . . . □ (11-5)
- an employee of a university . . . . □ (11-6)
- an employee of a government agency . . . . □ (11-7)
- employed in an unknown way . . . . □ (11-8)

u first obtained a **lead** to the information acquired in this instance rather than the **substance** of the informa-
itself, please answer question 7. If you directly obtained the information itself, skip question 7 and pro-
to question 8.

 first obtained the **lead** to this information

ORALLY FROM:

- a. an engineer or scientist employed by my own organization . . . . □ (12-1)
- b. an information scientist or librarian . . . . □ (12-2)
- c. a non-technical employee of my own organization (other than a librarian) . . . . □ (12-3)
- d. a sales representative of another company . . . . □ (12-4)
- e. a person not described above (please describe) _____
  _____ □ (12-5)

IN WRITTEN FORM IN:

- f. a report issued by my own organization . . . . □ (12-6)
- g. a bibliographical or abstract publication . . . . □ (12-7)
- h. a published book, technical or scientific journal, or conference preprint . . . . □ (12-8)
- i. a published trade magazine . . . . □ (12-9)
- j. a document not listed above (please describe) _____
  _____ □ (12-0)

Please describe the specific field from which this information came. The description of a field should sound
like the title of an advanced course that might be found at a university rather than a broad technical
discipline (e.g., Quantum Mechanics and Transistorized Circuit Theory rather than Physics and Electrical
Engineering): _____

Do you consider yourself experienced in this field? 1. Yes □ (13-1) 2. No □ (13-2)

Please describe the class of problem or application for which this information was useful. Once again,
please try to avoid describing broad problem areas such as Mechanical Design and Metallurgical Analysis.
Instead use phrases that would be appropriate to describe your work to a professional colleague (e.g.,
pump design or the interpretation of eutectic diagrams): _____

Do you consider yourself experienced with this type of problem? 1. Yes □ (14-1) 2. No □ (14-2)

The information was applied to a task which could be characterized as an example of the:

- a. formulation or testing of scientific theories, concepts, or models . . . . □ (15-1)
- b. empirical scientific investigation of physical phenomena . . . . □ (15-2)
- c. formulation, development, and investigation of new approaches to technical objectives . . . . □ (15-4)
- d. combination and integration of generally available designs and components into desired products,
  processes, and test procedures . . . . □ (15-6)
- e. refinement of existing products, processes, or test procedures . . . . □ (15-8)
- f. conduct of tests of materials, products, or processes . . . . □ (15-9)
- g. formulation, programming, or testing of approaches to data processing problems . . . . □ (15-0)
- h. following (please describe) _____

11. **The information obtained in this instance was primarily useful in that <u>stage</u> of technical or scientific work most closely identified with the:**

    a. generation of ideas to start or suggest a new project or program .... ☐ (

    b. preparation or evaluation of a proposal, plan, or outline of technical work to be done .... ☐ (

    c. investigation of a technical or scientific problem .... ☐ (

    d. design or execution of tests of theories, materials, or engineering designs .... ☐ (

    e. preparation or evaluation of technical reports .... ☐ (

    f. preparation or evaluation of standard specifications .... ☐ (

    g. following (please describe) _____

    _____ ☐ (

12. **Within this stage of work, the information was primarily useful in:**

    a. aiding in the perception and/or definition of a technical problem .... ☐ (

    b. choosing a data-gathering technique .... ☐ (

    c. choosing a plan or strategy for data collection .... ☐ (

    d. choosing a method of data analysis or interpretation .... ☐ (

    e. choosing a technical or scientific approach to the solution of a problem .... ☐ (

    f. aiding in the interpretation of collected data .... ☐ (

    g. (none of the above ways) .... ☐ (

13. **The primary manner in which the information affected my work was:**

    a. to suggest new approaches to the solution of a problem .... ☐ (

    b. to indicate the existence of new technical questions to be resolved .... ☐ (

    c. to change our confidence that a previously defined approach would yield expected results .... ☐ (

    d. to change our expectations on the costs and time needed to develop an approach to a problem .... ☐ (

    e. to alter the formal budget, schedule, or specifications associated with a problem .... ☐ (

    f. to affect the relations between a task and others .... ☐ (

    g. to certify known information or data .... ☐ (

    h. (in none of the above ways) .... ☐ (

PART II

BACKGROUND QUESTIONS

The following items ask you to describe some characteristics descriptive of you and your work. PLEASE CHECK THE <u>ONE</u> BOX MOST APPLICABLE FOR <u>EACH</u> ITEM.

14. **Job Type:**

    a. individual contributor .... ☐ (

    b. project head, group leader, supervisor, or other management or technical direction position .... ☐ (

    c. technical or scientific advisor or fellow .... ☐ (

    d. other (please describe): _____

    _____ ☐ (

15. **Kind of Organization in which you are employed:**

    a. a company primarily engaged in government contract work .... ☐ (

    b. a company primarily engaged in consumer goods manufacturing .... ☐ (

    c. a company primarily engaged in industrial goods manufacturing .... ☐ (

    d. a multidivision, diversified company .... ☐ (

    e. (another kind of profit-making enterprise) .... ☐ (

    f. a government agency .... ☐ (

    g. a not-for-profit or university-associated laboratory .... ☐ (

    h. an educational institution .... ☐ (

    i. (none of the above) .... ☐ (

Please re-examine your answer to question 15 above. If you checked categories a, b, c, d, or e (i.e., you are employ a company), please answer questions 16, 17, and 18. If you checked categories f, g, h, or i, skip questions 16, 17, ar and proceed to question 19.

Organizational unit in which you are employed:

a. research laboratory ☐ (22-1)
b. development department or laboratory ☐ (22-2)
c. engineering section or department ☐ (22-3)
d. sales or production department ☐ (22-4)
e. information processing or programing department ☐ (22-5)
f. (other staff group) ☐ (22-6)
g. (none of the above) ☐ (22-7)

Primary product of the organization:

a. missiles, aircraft or space systems ☐ (23-1)
b. computers ☐ (23-2)
c. controls systems or devices ☐ (23-3)
d. other electronically based systems ☐ (23-4)
e. electronic components ☐ (23-5)
f. chemical, petroleum, or related products ☐ (23-6)
g. primary metals or metal alloys ☐ (23-7)
h. machinery or machine systems ☐ (23-8)
i. other fabricated metal products ☐ (23-9)
j. (none of the above) ☐ (23-0)

Production process for the primary product of the organization:

a. small batch, unit, or custom production ☐ (24-1)
b. large batch or mass production ☐ (24-2)
c. continuing process production ☐ (24-3)
d. (none of the above) ☐ (24-4)

CEED TO QUESTION 19

I.E.E.E. Group Memberships:

a. industrial electronics and control instrumentation ☐ (25-1)
b. automatic control ☐ (25-2)
c. instrumentation and measurements ☐ (25-3)
d. circuit theory ☐ (25-4)
e. power ☐ (25-5)
f. component parts ☐ (25-6)
g. computer ☐ (25-7)
h. reliability ☐ (25-8)
i. electronic devices ☐ (25-9)
j. other groups (list) _____ (25-0)

20. Age:

a. 20–29 ☐ (26-1)
b. 30–39 ☐ (26-2)
c. 40–49 ☐ (26-3)
d. 50+ ☐ (26-4)

21. Employed by your present organization

a. for less than one year ☐ (30-1)
b. 1–2 years ☐ (30-2)
c. 3–5 years ☐ (30-3)
d. 6–10 years ☐ (30-4)
e. 11+ years ☐ (30-5)

22. Education: (check the highest level attained)

a. High School ☐ (32-1)
b. Certificate ☐ (32-2)
c. Bachelor's Degree ☐ (32-3)
d. Some Graduate Training ☐ (32-4)
e. Master's Degree ☐ (32-5)
f. Doctor's Degree ☐ (32-6)

23. Technical or scientific society meetings attended during the last year:

a. none ☐ (33-1)
b. one ☐ (33-2)
c. two ☐ (33-3)
d. three ☐ (33-4)
e. four or more ☐ (33-5)

24. Conference Papers, articles, or books published in the previous five years:

a. none ☐ (34-1)
b. one ☐ (34-2)
c. two to three ☐ (34-3)
d. four to six ☐ (34-4)
e. seven or more ☐ (34-5)

25. U.S. Patent applications filed during the previous five years:

a. none ☐ (35-1)
b. one ☐ (35-2)
c. two to three ☐ (35-3)
d. four to six ☐ (35-4)
e. seven to fifteen ☐ (35-5)
f. sixteen or more ☐ (35-6)

se complete the following sentence

6. I regularly read the following scientific, technical, and trade periodicals for the purpose of keeping up on my field: _____

_____

_____

(39)

The following seven items appear in the form of statements about people engaged in scientific or technical work in indu
Please indicate the strength of your agreement with each statement by checking the one box most applicable. Note
for the questions below the title professional refers to engineers, scientists, and computer programmers. If you are
engineer please read each statement as if it only pertained to engineers, if you are a scientist please read each staten
as if it only pertained to scientists.

27. Competent professionals should devote the large majority, if not all of their careers to their scientific
    or technical specialty.
    Strongly   Strongly
    a. Agree ☐ (40-1) b. Agree ☐ (40-2) c. Undecided ☐ (40-3) d. Disagree ☐ (40-4) e. Disagree ☐ (40-5)
28. Technical problems are generally more interesting than human or economic problems.
    Strongly   Strongly
    a. Agree ☐ (41-1) b. Agree ☐ (41-2) c. Undecided ☐ (41-3) d. Disagree ☐ (41-4) e. Disagree ☐ (41-5)
29. People with superior technical (or scientific) skills generally do not make the best associates.
    Strongly   Strongly
    a. Agree ☐ (42-1) b. Agree ☐ (42-2) c. Undecided ☐ (42-3) d. Disagree ☐ (42-4) e. Disagree ☐ (42-5)
30. Promotion upward and eventually out of technical (or scientific) ranks is the proper reward for su-
    perior work.
    Strongly   Strongly
    a. Agree ☐ (43-1) b. Agree ☐ (43-2) c. Undecided ☐ (43-3) d. Disagree ☐ (43-4) e. Disagree ☐ (43-5)
31. The more science (or technology) one knows the more he understands the world we all must live in.
    Strongly   Strongly
    a. Agree ☐ (44-1) b. Agree ☐ (44-2) c. Undecided ☐ (44-3) d. Disagree ☐ (44-4) e. Disagree ☐ (44-5)
32. Original technical (or scientific) ideas are generally among the least exciting events in corporate life.
    Strongly   Strongly
    a. Agree ☐ (45-1) b. Agree ☐ (45-2) c. Undecided ☐ (45-3) d. Disagree ☐ (45-4) e. Disagree ☐ (45-5)
33. A competent professional is so busy that it is unrealistic to expect him to set aside more than 5% of
    his time for extending and developing his technical (or scientific) knowledge and skill.
    Strongly   Strongly
    a. Agree ☐ (46-1) b. Agree ☐ (46-2) c. Undecided ☐ (46-3) d. Disagree ☐ (46-4) e. Disagree ☐ (46-5)

Please think of three people each of whom is a person you respect and would like to be respected by. For each of the t
people, as indicated by the column below marked "1 − 2 − 3," check one box corresponding to the most applicable
scription.

	1	2	
a manager or supervisor of professional activities in your own organization	☐	☐	▪
another professional in your own organization	☐	☐	▪
a professionally employed person in another organization	☐	☐	▪
a person whose activities are not in science, engineering, or data processing	☐	☐	▪
	(47)	(48)	(∙

My profession is that of a:

Programmer .... ☐ (50-1)
Electrical Engineer .... ☐ (50-2)
Mechanical Engineer .... ☐ (50-3)
Metallurgical Engineer .... ☐ (50-4)
Other Engineer .... ☐ (50-5)
Chemist .... ☐ (50-6)
Mathematician .... ☐ (50-7)
Physicist .... ☐ (50-8)
Metallurgist .... ☐ (50-9)
Other Scientist .... ☐ (50-0)

TYPICAL COVER LETTER

Corporate Surveys

SUBJECT: Technical Information Flow:

The attached questionnaire is concerned with the nature, sources, and use of the technical information which forms a part of your everyday work. The questionnaire was prepared by an academic research team from the Harvard Business School working under a grant from the National Science Foundation.

I urge you, during the 15 to 20 minutes required to fulfill this request, to treat this piece of research with the same respect you would accord to research or design work in your own field. We expect that this study will contribute to a better understanding of the patterns of communication of technical information in general, and we hope it will give us information which will help us devise ways to improve our technical communications in this company.

The questionnaire was designed to safeguard both the Company's proprietary information and your individual interests. Your answers will be analyzed by the research team and reported only in aggregate statistical terms. Please complete the questionnaire at one sitting and return it at the time requested.

Thank you for your important contribution to this effort.

John Doe
Advanced Development Division
General Manager

PART II

BACKGROUND QUESTIONS

16. Name: _____

17. Job Title: _____   (2▌

The following items ask you to describe some characteristics descriptive of you and your work. Please check the <u>one</u> box most applicable for <u>each</u> item.

18. Age:

20–29	☐	(26-1)
30–39	☐	(26-2)
40–49	☐	(26-3)
50+	☐	(26-4)

19. Employed by _____ for:

less than one year	☐	(30-1)
1–2 years	☐	(30-2)
3–5 years	☐	(30-3)
6–10 years	☐	(30-4)
11+ years	☐	(30-5)

20. Profession:

Programmer	☐	(31-1)
Electrical Engineer	☐	(31-2)
Mechanical Engineer	☐	(31-3)
Metallurgical Engineer	☐	(31-4)
Other Engineer	☐	(31-5)
Chemist	☐	(31-6)
Mathematician	☐	(31-7)
Physicist	☐	(31-8)
Metallurgist	☐	(31-9)
Other Scientist	☐	(31-0)

21. Education: (check the highest level attained)

High School	☐	(32-1)
Certificate	☐	(32-2)
Bachelor's Degree	☐	(32-3)
Some Graduate Training	☐	(32-4)
Master's Degree	☐	(32-5)
Doctor's Degree	☐	(32-6)

22. Technical or scientific society meetings attended in the last year:

none	☐	(33-1▌
one	☐	(33-2▌
two	☐	(33-3▌
three	☐	(33-4▌
four or more	☐	(33-5▌

23. Papers, articles, or books published in the previous five years:

none	☐	(34-1▌
one	☐	(34-2▌
two to three	☐	(34-3
four to six	☐	(34-4
seven or more	☐	(34-5

24. U.S. Patents filed during the previous five years:

none	☐	(35-1
one	☐	(35-2
two to three	☐	(35-3
four to six	☐	(35-4
seven to fifteen	☐	(35-5
sixteen or more	☐	(35-6

25. "Outstanding Technical Contribution Awards" won in the previous five years:

none	☐	(36-1
one	☐	(36-2
two or more	☐	(36-3▌

Please complete the following sentence

26. I regularly read the following scientific, technical, and trade periodicals for the purpose of keeping up on my field: _____

_____

_____   (39▌

(CONTINUED ON THE NEXT PAGE

The following seven items appear in the form of statements about people engaged in scientific or technical work in industry. Please indicate the strength of your agreement with each statement by checking the one box most applicable. Note that for the questions below the title professional refers to engineers, scientists, and computer programmers. If you are an engineer please read each statement as if it only pertained to engineers, if you are a scientist please read each statement as if it only pertained to scientists. If you are a programmer, please read each statement as if it only pertained to programmers.

27. Competent professionals should devote the large majority, if not all of their careers to their scientific or technical specialty.
    Strongly   Strongly
    a. Agree ☐ (40-1) b. Agree ☐ (40-2) c. Undecided ☐ (40-3) d. Disagree ☐ (40-4) e. Disagree ☐ (40-5)

28. Technical problems are generally more interesting than human or economic problems.
    Strongly   Strongly
    a. Agree ☐ (41-1) b. Agree ☐ (41-2) c. Undecided ☐ (41-3) d. Disagree ☐ (41-4) e. Disagree ☐ (41-5)

29. People with superior technical (or scientific) skills generally do not make the best associates.
    Strongly   Strongly
    a. Agree ☐ (42-1) b. Agree ☐ (42-2) c. Undecided ☐ (42-3) d. Disagree ☐ (42-4) e. Disagree ☐ (42-5)

30. Promotion upward and eventually out of technical (or scientific) ranks is the proper reward for superior work.
    Strongly   Strongly
    a. Agree ☐ (43-1) b. Agree ☐ (43-2) c. Undecided ☐ (43-3) d. Disagree ☐ (43-4) e. Disagree ☐ (43-5)

31. The more science (or technology) one knows the more he understands the world we all must live in.
    Strongly   Strongly
    a. Agree ☐ (44-1) b. Agree ☐ (44-2) c. Undecided ☐ (44-3) d. Disagree ☐ (44-4) e. Disagree ☐ (44-5)

32. Original technical (or scientific) ideas are generally among the least exciting events in corporate life.
    Strongly   Strongly
    a. Agree ☐ (45-1) b. Agree ☐ (45-2) c. Undecided ☐ (45-3) d. Disagree ☐ (45-4) e. Disagree ☐ (45-5)

33. A competent professional is so busy that it is unrealistic to expect him to set aside more than 5% of his time for extending and developing his technical (or scientific) knowledge and skill.
    Strongly   Strongly
    a. Agree ☐ (46-1) b. Agree ☐ (46-2) c. Undecided ☐ (46-3) d. Disagree ☐ (46-4) e. Disagree ☐ (46-5)

Please think of three people each of whom is a person who stimulates your work or thought, whose judgment you respect and whose recognition of your accomplishments would be valuable to you. For each of the three people, as indicated by the column below marked "1 — 2 — 3," check one box corresponding to the most applicable description.

	1	2	3
a manager or supervisor of professional activities at DATA	☐	☐	☐
another professional in your own division of DATA	☐	☐	☐
another professional in some other division of DATA	☐	☐	☐
a professionally employed person in another organization	☐	☐	☐
a person whose activities are not in science, engineering, or data processing	☐	☐	☐
	(47)	(48)	(49)

Please circle the one scale value which most describes the extent to which you think of yourself as a SPECIALIST (relative to colleagues in your own division) in a particular or limited set of discipline areas (i.e., Transistor Circuit Theory, Boolean Algebra, Vibration Analysis, etc.):

Not at all   Somewhat   Very much

1   2   3   4   5   6   7   (5

Please circle the one scale value which most describes the extent to which you think of yourself as a SPECIALIST (relative to colleagues in your own division) in a particular or limited set of application areas (i.e., Amplifier Design, Random Number Generators, Printer Development, etc.):

Not at all   Somewhat   Very much

1   2   3   4   5   6   7   (5

Appendix B

Notes on the Survey Data

The survey data used in our analysis are substantially valid and reliable, and the principal assumptions required for the inferences drawn from that analysis are essentially sound. These conclusions are based on empirical evidence, obtained either through internal analysis of the data or through field work designed to test the survey instruments. This appendix will present the evidence and details of analysis that bear on an evaluation of the amount of error variance in the survey data and the stability of these data over time.

Design of the Questionnaire

The matter of choosing a data-gathering method involves trading off limitations and advantages. We decided on the use of a self-administered questionnaire because we believe it is essential to have a large sample base in order to unravel a complex multivariable process. However, this decision placed upon us a responsibility for exact, unambiguous wording of the questionnaire. The questionnaire used in the survey is presented as Appendix A.

One important aid to designing a self-administered questionnaire is an extensive amount of presurvey field testing. The questionnaire used in this study is a modification of the questionnaire used in a previous study.[1] That questionnaire was refined in its design and subse-

---

[1] C. P. McLaughlin, R. S. Rosenbloom, and F. W. Wolek, "Technology Transfer and the Flow of Technical Information in a Large Industrial Corporation," 1965.

quently validated in its interpretation through 12 presurvey and more than 30 postsurvey interviews. Presurvey and postsurvey interviews with executives of establishments participating in both studies also assured us of the relevance of the wording used in each instrument and the general validity of the results obtained. After the questionnaire was redesigned for the present study, it was tested again in 6 presurvey interviews and 21 postsurvey interviews in 5 different establishments, and in over 50 telephone interviews.

In the test interviews each subject gave, and where necessary explained, his answers to a questionnaire. In the postsurvey tests, the subjects were respondents to the survey itself; the interview, however, was primarily based on the step-by-step completion of a second questionnaire. These interviews showed that in general the respondents had little difficulty understanding and answering the questions used.

The validity of the data collected in the personal background section (Part II) of the questionnaire is generally self-evident. The only difficulty arose in Question 25, in which the phrase "regularly read" (periodicals) was somewhat ambiguous. Nevertheless, the correlation of this item with other related items (i.e., education, conferences attended, and papers published) suggests that the data are generally valid.

More serious questions arise in connection with Part I of the questionnaire. Even though the typical respondent understood the import of the survey questions, he may still have found some difficulty in fitting a complex, personal experience into the framework given. This situation, of course, grows out of the difficulty of trying to separate the interacting elements of a process that tends to be not only viewed but acted out as a whole.

In some cases, for example, the information may have been obtained simultaneously in more than one way, as, for example, when a salesman discusses a technical problem and also supplies some literature on the same subject. One consequence of this problem is that it is likely that some respondents failed to identify on Question 7 a lead to their source of information (i.e., the aggregate occurrence of leads in our data is not fully valid). A second consequence of the problem is that a number of questionnaires (generally only 5%–10% of the cases for any one establishment) indicated both interpersonal and written media in answer to Question 5. Responses indicating the use of multiple sources were eliminated from the analysis.

Our interviews suggest that the validity of some data on the function of information (i.e., data from Questions 11, 12, and 13) is particularly doubtful, and these data have been excluded from our analysis. On the other hand, our field interviews also indicate that respondents had no difficulty with Question 10 and that the responses are valid indications of the kind of task to which the information was applied.

The difficulties with Questions 11 through 13 are of three kinds:

(a) Interpretation problems: Categories b, c, and d of Question 12 were intended to enable us to distinguish information dealing with methodology; these categories, however, were not always interpreted this way (e.g., several respondents thought they dealt with computer processing).

(b) Structural problems: Field data indicate that the categories used in Question 11 and perhaps in Question 12 are not mutually exclusive (e.g., in many cases investigation of a problem has gone into great depth during the planning phases). Furthermore, there is some reason to believe that some respondents gave a substantive answer when the correct reply was "none of the above."

(c) Conceptual problems: The respondents often had difficulty distinguishing problems from projects and projects from programs (e.g., in one case information on a problem of test-site preparation was described as 11(c) rather than 11(d) because the test was being run to investigate a problem).

The Incident-Selection Process

By far the most significant question bearing on the validity of the questionnaire concerns the incident chosen for description. We were very much interested in learning whether the respondents were biased toward selecting particular kinds of incidents. Field experience in the previous survey suggested that several respondents searched for an instance that stood out in their minds rather than the most recent instance. For this reason we were careful to emphasize the importance of the instruction "most recent" in the present survey. Field studies suggest that this instruction was taken seriously and was followed, barring other difficulties described below. We believe that some respondents selected incidents that, though recent, also stood out in their minds (as some respondents noted, "one that I knew I could answer the questions on"), but we suggest that this tendency may have improved the validity of the responses, since the incidents thus selected were ones that more clearly involved definite use of the information.

Field interviews further suggest some slight problems with the instruction that the information should have been "useful in work." In the majority of cases the respondent did choose an incident in which information was directly applied to a specific task. In a few cases, however, the chosen incidents might be described as fitting a transition zone between direct application and competence development. This transition zone exists because the competences people attempt to develop are generally related to foreseen work; as a project gradually emerges into current work, this distinction between direct application and competence development blurs.

In general, we are convinced that there is little obvious bias in the selection of incidents. However, we cannot fail to note a general concern about many cases in which respondents in an interview expressed extreme discomfort over choosing an incident. We have come to believe that this discomfort existed because some respondents found it difficult to focus on a specific incident given only a general definition of "information."

Validity of the Scale of Professional Orientation

The analysis reported in Chapter 4 is based, in part, on a measure of the degree of a respondent's professional orientation. The measure of this personal orientation combines an attitudinal index of "professional identity" and an index of past behavior indicative of a "commitment to the development of professional competence." There are several sources of error variance in the data including problems in definition, item selection, item phrasing, and item weighting. A program of field work on these sources of error included studies of (a) 26 respondents to the survey who filled out an additional questionnaire and were interviewed on their responses and their general approach to work, and (b) 58 engineers and scientists, in three organizations not included in our survey, who provided scores on the measure of professional orientation, a personal evaluation of their own professional orientation, and peer evaluations of professional orientation.

This program of field work led us to conclude that (a) the index used does measure a personal orientation substantially like the professional orientation defined in Chapter 4, (b) the sources of error are sufficiently diverse and random in effect that no significant interpretation other than one based on professional orientation can reasonably be ascribed to the relationships obtained, and (c) the total error variance is high, particularly for the attitude items.

The construct of professional orientation is a general one. That is, it is a pattern of attitudes and behavior that is conceptualized at a high level of abstraction. Given such a construct, it is impossible to obtain a truly "noiseless" measure.[2] The best procedure in such a case is to limit reliance on any one factor or small set of factors and to average the responses over a number of different indices. This procedure for scale construction is called the Likert Technique, and it was followed in this study.[3]

A particular source of error in the measure of professional identity lies in the phrasing of the attitude items used. The proper balance between ambiguity and clarity in phrasing is difficult to determine on other than empirical grounds. The danger of respondent bias in attempting to "paint a good picture" balances the danger of a high error due to ambiguity. In the present case, interview data suggest that our attempt to find a good balance was not successful. The questions proved to contain more ambiguity than was necessary. However, the strength of value judgments generally surrounding this construct was also evident in the interviews.

Another consideration in phrasing items involves the distinction between a general and a personal reference (e.g., "promotion out of professional ranks is the best reward for superior work" as opposed to "I would prefer promotion out of professional ranks"). Our interviews suggest that the use of the former approach was unnecessary and led to the introduction of significant error variance. Many indicated that their replies on some items would have been different (often dramatically so) had another point of reference been used.

Computational procedures are also important in the construction of any index. Given the high error variance in several items (particularly the attitudinal measures), the limitations of the Likert procedure are inconsequential, and its computational advantages significant. Thus a simple average of answered items was computed for each subject.

One concern in a Likert scale is the location of the "zero point" in the scoring for each item. Since the wording of the items is arbitrary, there is some question as to whether an "agree" answer on one item has the same meaning as an "agree" on another. A method of resolving this difficulty is to find the modal response for each item and to use

---

[2] This fact was clearly demonstrated in a study of peer judgments of professional orientation versus scale measurement. Within a relatively small group of 20 people, the list of persons nominated by their peers as the most professionally oriented included well over half the group and showed only minimal agreement.

[3] The selection of a scaling procedure was guided by several sources, for example, Ghiselli, *Theory of Psychological Measurement*, 1964; and L. Festinger and D. Katz, *Research Methods in the Behavioral Sciences*, 1953.

that response as the zero point on the item. When this analysis was carried out, a rank-order correlation analysis of the corrected and uncorrected versions gave a .95 correlation coefficient, indicating that the correction added little. Hence this correction procedure was not employed in computing the index we used.

Field interview studies also were useful in indicating that several possible sources of error were not generally important. For example, on the question of respondent attitude toward the instrument, the interviews suggest that while some error was introduced by a perceived provocativeness in wording, the general attitude of respondents was one of mild but serious interest. Furthermore, in only one or two instances did a respondent demonstrate any bias in trying to "paint a good picture" of himself. Nevertheless, field work did indicate that strong value connotations are attached to a professional orientation. Sometimes these were opposing but generally they favored a professional orientation as "good." In one study many respondents refused or otherwise failed to nominate anyone as having a low professional orientation relative to his group. The modest success of the instrument in avoiding strong effects from these value judgments probably is related to that same ambiguity in wording which introduced significant error variance in its own right.

Field data and other tests were also useful in indicating that two particular items (30 and 32 in Part II of the questionnaire) had a sufficiently inconsistent response to justify eliminating them in the scoring procedure. Finally, the field program suggests that little error was introduced by organizational climate. In other words, respondents in laboratories tended to reflect their personal beliefs and not those of the organizational colleagues. Indeed, the interviewer's impressions suggest that a less "noisy" measure might find that these institutions were considerably less polarized than is commonly portrayed.

Stability Over Time

Two special surveys, separated by six months in time, offer an opportunity to consider the reliability of measures taken at only one point in time.[4] Analysis of these samples shows that if we were to take either of the surveys by itself as a basis for estimating the pattern

---

[4] These surveys were part of an overall study of information flow to members of the Institute of Electrical and Electronics Engineers (IEEE). As is clear in Table B-1, this more professionally oriented group of respondents follows a pattern of acquisition which differs from that found for the average engineer in our corporate sample.

of information flow, we would not get substantially different answers from those obtained by the method we have used, that is, combining data from the two surveys. Additional evidence on this point is a comparison of data from the first survey with a sub-sample of respondents in the second survey who had been included in both. Although we would suspect that the population of those who would be willing to respond a second time is probably not representative of all those who responded to the first survey, even this factor is not enough to introduce a meaningful difference in the data for these two groups (see Table B-1). Thus it seems safe to conclude that any error introduced by the tactic of sampling the population simultaneously at only one point in time is small relative to the other sources of error in our estimates and may be disregarded.

While the data presented in Table B-1 support the reliability of our

TABLE B-1

Reliability Analysis

	Respondent Group		
Source:	First Survey	Second Survey Only	Both Surveys
Organizational Interpersonal	14.9%	20.5%	13.1%
External-Interpersonal	25.5	27.0	32.1
Organizational Documents	6.0	4.6	6.0
Trade	26.8	24.9	21.4
Professional	26.8	23.0	27.4
	100.0%	100.0%	100.0%
	N = 497	N = 370	N = 84

measures over time, we call attention to a different effect of time on our survey data. The instances reported in our surveys were consistently very recent — 40% within two days, 60% within a week, 80% within three weeks. However, as shown in Table B-2, after an interval of three weeks there is a noticeable bias toward recollection of formal literature and away from local, interpersonal, or informal means. This, of course, has important implications for the use of retrospective descriptions of incidents when the selection method does not yield very recent events.

TABLE B-2

Effect of Time Interval
(*IEEE Survey Data*)

Source:	Interval (in days)		
	0–6	7–20	Over 20
Local and other company Interpersonal	18.2%	14.3%	9.8%
External-Interpersonal	25.8	25.0	19.6
Company documents	5.7	5.8	3.4
Trade	22.8	24.1	26.0
Professional	21.3	21.9	28.4
Mixed (i.e., both oral and written source)	6.1	8.9	12.7
	100%	100%	100%
	N = 685	N = 224	N = 204

(The probability of differences this large occurring by chance is less than .01.)

Appendix C

Background Data

The following tables summarize selected personal characteristics of the survey respondents. For the corporate surveys these are given by place of employment and by professional field; IEEE data are tabulated according to group membership. The classification of employing organizations is that used throughout the report; the units are described in Table 2-1. Distinctions as to profession are made according to the respondent's answer to Question 20 in Part II of the questionnaire. Table C-1 tabulates responses by place of employment and professional field. The IEEE tabulations here exclude respondents who indicated membership in more than one of the four groups listed.

List of Tables

Table C-1

Professional Field by Place of Employment
(Surveys in Corporations)

	MEDCO	BASIC	POLY	MECHO	CONTROLS	EDP	EQUIPMENT	ADVANCED DEVELOPMENT	RESEARCH
Life Scientists	120 0.538	0 0.	0 0.	0 0.	0 0.	0 0.	0 0.	0 0.	0 0.
Chemists	80 0.359	114 0.655	86 0.705	11 0.066	1 0.005	0 0.	5 0.019	1 0.007	37 0.117
All Other Scientists	9 0.040	12 0.069	2 0.016	6 0.036	1 0.005	21 0.094	15 0.056	36 0.243	151 0.479
Electrical Engineers	1 0.004	1 0.006	0 0.	24 0.145	178 0.813	91 0.406	172 0.642	59 0.399	66 0.210
Mechanical Engineers	0 0.	2 0.011	2 0.016	86 0.518	33 0.151	0 0.	62 0.231	16 0.068	3 0.010
All Other Engineers	13 0.058	45 0.259	32 0.262	39 0.235	6 0.027	29 0.129	12 0.045	16 0.108	11 0.035
Computer Programmers	0 0.	0 0.	0 0.	0 0.	0 0.	83 0.371	2 0.007	26 0.176	47 0.149
	N= 223	N= 174	N= 122	N= 166	N= 219	N= 224	N= 268	N= 148	N= 315

Table C-2

Age Within Various Companies

Age:	MEDCO	BASIC	POLY	MECHO	CONTROLS	EDP	EQUIPMENT	ADVANCED DEVELOPMENT	RESEARCH
20–29	46 0.188	38 0.209	31 0.248	37 0.220	65 0.295	66 0.289	78 0.304	19 0.164	56 0.190
30–39	91 0.371	66 0.363	53 0.424	59 0.351	78 0.355	125 0.548	108 0.420	62 0.534	172 0.583
40–49	81 0.331	56 0.308	35 0.280	59 0.351	59 0.268	30 0.132	51 0.198	32 0.276	66 0.224
50+	27 0.110	22 0.121	6 0.048	13 0.077	18 0.082	7 0.031	20 0.078	3 0.026	1 0.003
	M= 245	M= 182	M= 125	M= 168	M= 220	M= 228	M= 257	M= 116	M= 295

TABLE C-3

Seniority Within Various Companies

Seniority in Years:	MEDCO	BASIC	POLY	MECHO	CONTROLS	EDP	EQUIPMENT	ADVANCED DEVELOPMENT	RESEARCH
Under 1	20 0.082	13 0.072	10 0.080	12 0.071	28 0.127	15 0.066	9 0.033	9 0.060	39 0.122
1-2	38 0.155	25 0.138	24 0.192	12 0.071	32 0.145	43 0.189	46 0.170	21 0.141	51 0.160
3-5	49 0.200	45 0.249	33 0.264	27 0.161	44 0.200	76 0.333	63 0.233	46 0.309	102 0.320
6-10	38 0.155	54 0.298	22 0.176	27 0.161	33 0.150	59 0.259	65 0.241	53 0.356	101 0.317
11+	100 0.408	44 0.243	36 0.288	90 0.536	83 0.377	35 0.154	87 0.322	20 0.134	26 0.082
	M= 245	M= 181	M= 125	M= 168	M= 220	M= 228	M= 270	M= 149	M= 319

Table C-4
Education of Various Company Personnel

Education:	MEDCO	BASIC	POLY	MECHO	CONTROLS	EDP	EQUIPMENT	ADVANCED DEVELOPMENT	RESEARCH
High School Only	5 0.020	6 0.033	3 0.024	2 0.012	1 0.005	19 0.084	17 0.063	5 0.034	3 0.009
Certificate	8 0.033	10 0.055	4 0.032	1 0.006	3 0.014	13 0.057	16 0.059	9 0.060	2 0.006
Bachelors	52 0.212	32 0.177	27 0.216	58 0.345	136 0.618	78 0.344	86 0.320	22 0.148	29 0.091
Graduate Study	34 0.139	31 0.171	30 0.240	54 0.321	66 0.300	84 0.370	76 0.283	31 0.208	38 0.119
Masters	43 0.176	47 0.260	21 0.168	44 0.262	14 0.064	30 0.132	60 0.223	73 0.490	77 0.241
Doctors	103 0.420	55 0.304	40 0.320	9 0.054	0 0.	3 0.013	14 0.052	9 0.060	170 0.533
	M= 245	M= 181	M= 125	M= 168	M= 220	M= 227	M= 269	M= 149	M= 319

TABLE C-5

Meetings Attended by Various Company Personnel

Meetings Attended (Past Year):	MEDCO	BASIC	POLY	MECHO	CONTROLS	EDP	EQUIPMENT	ADVANCED DEVELOP- MENT	RESEARCH
None	39	7	30	52	109	95	156	51	43
	0.159	0.206	0.240	0.310	0.498	0.419	0.584	0.342	0.135
One	57	15	40	33	50	61	64	47	78
	0.233	0.441	0.320	0.196	0.228	0.269	0.240	0.315	0.244
Two	56	4	22	31	27	34	25	31	103
	0.229	0.118	0.176	0.185	0.123	0.150	0.094	0.208	0.323
Three	32	4	19	20	12	13	9	12	53
	0.131	0.118	0.152	0.119	0.055	0.057	0.034	0.081	0.166
Four or More	61	4	14	32	21	24	13	8	42
	0.249	0.118	0.112	0.190	0.096	0.106	0.049	0.054	0.132
	M= 245	M= 34	M= 125	M= 168	M= 219	M= 227	M= 267	M= 149	M= 319

TABLE C-6

Publications by Various Company Personnel

Papers Published (5 Years):	MEDCO	BASIC	POLY	MECHO	CONTROLS	EDP	EQUIPMENT	ADVANCED DEVELOP- MENT	RESEARCH
None	8C 0.329	108 0.607	78 0.650	107 0.648	170 0.776	172 0.758	187 0.706	73 0.493	63 0.198
One	41 0.169	19 0.107	20 0.167	26 0.158	33 0.151	16 0.070	34 0.128	33 0.223	37 0.116
2-3	57 0.235	37 0.208	16 0.133	19 0.115	15 0.068	31 0.137	35 0.132	23 0.155	78 0.245
4-6	4C 0.165	8 0.045	6 0.050	7 0.042	1 0.005	7 0.031	7 0.026	13 0.088	56 0.176
Seven or More	25 0.103	6 0.034	0 0.	6 0.036	0 0.	1 0.004	2 C.008	6 0.041	84 0.264
	M= 243	M= 178	M= 120	M= 165	M= 219	M= 227	M= 265	M= 148	M= 318

Table C-7

Patents Filed by Various Company Personnel

Patents Filed (5 Years):	MEDCO	BASIC	POLY	MECHO	CONTROLS	EDP	EQUIPMENT	ADVANCED DEVELOPMENT	RESEARCH
None	160 0.672	99 0.559	55 0.447	129 0.782	164 0.752	183 0.813	188 0.709	102 0.694	194 0.610
One	24 0.101	25 0.141	22 0.179	18 0.109	30 0.138	18 0.080	35 0.132	18 0.122	42 0.132
2-3	28 0.118	36 0.203	15 0.122	11 0.067	17 0.078	17 0.076	30 0.113	11 0.075	54 0.170
4-6	11 0.046	14 0.079	17 0.138	4 0.024	4 0.018	6 0.027	11 0.042	9 0.061	18 0.057
7-15	12 0.050	3 0.017	7 0.057	2 0.012	3 0.014	1 0.004	1 0.004	6 0.041	8 0.025
Sixteen or More	3 0.013	0 0.	7 0.057	1 0.006	0 0.	0 0.	0 0.	1 0.007	2 0.006
	M= 238	M= 177	M= 123	M= 165	M= 218	M= 225	M= 265	M= 147	M= 318

Table C-8

Periodicals Read by Various Company Personnel

Periodicals Read Regularly:	MEDCO	BASIC	POLY	MECHO	CONTROLS	EDP	EQUIPMENT	ADVANCED DEVELOPMENT	RESEARCH
None	5 0.022	4 0.024	0 0.	5 0.033	4 0.020	10 0.047	11 0.043	2 0.014	11 0.036
One	13 0.057	8 0.048	4 0.033	8 0.053	25 0.124	37 0.174	35 0.136	9 0.063	14 0.046
Two	23 0.100	19 0.114	14 0.115	23 0.153	44 0.219	44 0.207	57 0.221	15 0.106	35 0.115
3-5	82 0.357	76 0.458	52 0.426	63 0.420	102 0.507	93 0.437	105 0.407	70 0.493	131 0.431
Six or More	107 0.465	59 0.355	52 0.426	51 0.340	26 0.129	29 0.136	50 0.194	46 0.324	113 0.372
M=	230	166	122	150	201	213	258	142	304

TABLE C-9

Age Within Various Professions

Age:	Life Scientists	Chemists	*Computer Scientists	Electri-cal Engineers	Mechani-cal Engineers	Other Engi-neers	Programmers
20-29	27 0.225	53 0.160	31 0.154	165 0.288	47 0.240	41 0.206	135 0.396
30-39	40 0.333	139 0.420	122 0.607	262 0.457	68 0.347	87 0.437	171 0.501
40-49	42 0.350	108 0.326	46 0.229	121 0.211	60 0.306	54 0.271	31 0.091
50+	11 0.092	31 0.094	2 0.010	25 0.044	21 0.107	17 0.085	4 0.012
	M= 120	M= 331	M= 201	M= 573	M= 196	M= 199	M= 341

*This category includes only those mathematicians and physicists employed in data processing businesses, excluding others classified as "other scientists" in Table C-1.

Table C-10

Seniority in the Professional Field

Seniority in Years:	Life Scientists	Chemists	Computer Scientists	Electrical Engineers	Mechanical Engineers	Other Engineers	Programmers
Under 1	13 0.108	19 0.057	19 0.084	44 0.074	14 0.070	17 0.083	38 0.094
1-2	21 0.175	41 0.122	37 0.164	92 0.155	23 0.116	28 0.136	92 0.228
3-5	23 0.192	90 0.269	75 0.333	149 0.251	38 0.191	43 0.209	138 0.342
6-10	23 0.192	73 0.218	78 0.347	142 0.239	39 0.196	50 0.243	115 0.285
11+	40 0.333	112 0.334	16 0.071	167 0.281	85 0.427	68 0.330	20 0.050
	M= 120	M= 335	M= 225	M= 594	M= 199	M= 206	M= 403

TABLE C-11

Education of Professional Personnel

Education:	Life Scientists	Chemists	Computer Scientists	Electrical Engineers	Mechanical Engineers	Other Engineers	Programmers
High School Only	2 0.017	1 0.003	3 0.013	14 0.024	7 0.035	9 0.044	38 0.094
Certificate	3 0.025	6 0.018	4 0.018	17 0.029	8 0.040	12 0.058	15 0.037
Bachelors	22 0.183	60 0.180	18 0.080	225 0.380	85 0.427	58 0.282	155 0.385
Graduate Study	15 0.125	66 0.198	21 0.093	188 0.318	54 0.271	44 0.214	105 0.261
Masters	17 0.142	68 0.204	56 0.249	113 0.191	40 0.201	63 0.306	81 0.201
Doctors	61 0.508	133 0.398	123 0.547	35 0.059	5 0.025	20 0.097	9 0.022
	M= 120	M= 334	M= 225	M= 592	M= 199	M= 206	M= 403

TABLE C-12

Meetings Attended by Professional Personnel

Meetings Attended (Past Year):	Life Scientists	Chemists	Computer Scientists	Electrical Engineers	Mechanical Engineers	Other Engineers	Programmers
None	22 0.183	39 0.172	32 0.142	255 0.431	100 0.508	52 0.291	231 0.575
One	29 0.242	73 0.322	47 0.209	149 0.252	35 0.178	51 0.285	97 0.241
Two	22 0.183	49 0.216	70 0.311	101 0.171	27 0.137	29 0.162	46 0.114
Three	16 0.133	30 0.132	45 0.200	38 0.064	15 0.076	19 0.106	6 0.015
Four or More	31 0.258	36 0.159	31 0.138	48 0.081	20 0.102	28 0.156	22 0.055
	M= 120	M= 227	M= 225	M= 591	M= 197	M= 179	M= 402

TABLE C-13

Publications by Those in the Professional Field

Papers Published (5 Years):	Life Scientists	Chemists	Computer Scientists	Electrical Engineers	Mechanical Engineers	Other Engineers	Programmers
None	36 / 0.300	145 / 0.448	42 / 0.187	390 / 0.662	145 / 0.740	134 / 0.657	306 / 0.761
One	16 / 0.133	54 / 0.167	22 / 0.098	87 / 0.148	26 / 0.133	27 / 0.132	43 / 0.107
2-3	30 / 0.250	66 / 0.204	54 / 0.241	79 / 0.134	20 / 0.102	29 / 0.142	44 / 0.109
4-6	23 / 0.192	30 / 0.093	40 / 0.179	23 / 0.039	4 / 0.020	12 / 0.059	6 / 0.015
Seven or More	15 / 0.125	29 / 0.090	66 / 0.295	10 / 0.017	1 / 0.005	2 / 0.010	3 / 0.007
	M= 120	M= 324	M= 224	M= 589	M= 196	M= 204	M= 402

TABLE C-14

Patents Filed by Professional Personnel

Patents Filed (5 Years):	Life Scientists	Chemists	Computer Scientists	Electrical Engineers	Mechanical Engineers	Other Engineers	Programmers
None	91 0.778	155 0.473	153 0.680	393 0.671	145 0.740	145 0.714	386 0.965
One	10 0.085	53 0.162	27 0.120	82 0.140	21 0.107	25 0.123	11 0.027
2-3	6 0.051	59 0.180	30 0.133	78 0.133	16 0.082	21 0.103	3 0.007
4-6	4 0.034	33 0.101	10 0.044	26 0.044	8 0.041	10 0.049	0 0.
7-15	6 0.051	17 0.052	4 0.018	7 0.012	5 0.026	1 0.005	0 0.
Sixteen or More	0 0.	11 0.034	1 0.004	0 0.	1 0.005	1 0.005	0 0.
	M= 117	M= 328	M= 225	M= 586	M= 196	M= 203	M= 400

TABLE C-15

Periodicals Read by Professional Personnel

Periodicals Read Regularly:	Life Scientists	Chemists	Computer Scientists	Electrical Engineers	Mechanical Engineers	Other Engineers	Programmers
None	3 0.028	4 0.C12	4 0.019	17 0.031	7 0.038	5 0.026	36 0.101
One	7 0.064	14 0.043	7 0.033	59 0.106	21 0.114	5 0.026	83 0.232
Two	11 0.101	36 0.112	17 0.080	112 0.201	34 0.184	29 0.151	102 0.286
3-5	26 0.239	137 0.425	95 0.446	258 0.463	86 0.465	92 0.479	116 0.325
Six or More	62 0.569	131 0.407	90 0.423	111 0.199	37 0.200	61 0.318	20 0.056
	N= 109	N= 322	N= 213	N= 557	N= 185	N= 192	N= 357

TABLE C-16

Age Within IEEE Group

Age:	IECI*		Circuit Theory		Power		Computer	
20-29	5	0.093	40	0.278	27	0.084	82	0.231
30-39	22	0.407	71	0.493	81	0.252	185	0.521
40-49	16	0.296	24	0.167	111	0.345	74	0.208
50+	11	0.204	9	0.062	103	0.320	14	0.039
	M= 54		M= 144		M= 322		M= 355	

TABLE C-17

Seniority Within IEEE Group

Seniority in Years:	IECI		Circuit Theory		Power		Computer	
Under 1	4	0.074	15	0.104	14	0.043	44	0.124
1-2	10	0.185	27	0.187	21	0.065	73	0.206
3-5	13	0.241	48	0.333	28	0.087	96	0.270
6-10	7	0.130	31	0.215	52	0.161	77	0.217
11+	20	0.370	23	0.160	207	0.643	65	0.183
	M= 54		M= 144		M= 322		M= 355	

*IECI refers to Industrial Electronics and Control Instrumentation.

Table C-18
Education Within IEEE Group

Education:	IECI	Circuit Theory	Power	Computer
High School Only	0 / 0.	0 / 0.	2 / 0.006	3 / 0.008
Certificate	0 / 0.	2 / 0.014	10 / 0.031	4 / 0.011
Bachelors	17 / 0.315	19 / 0.132	146 / 0.453	55 / 0.155
Graduate Study	19 / 0.352	48 / 0.333	118 / 0.366	122 / 0.344
Masters	15 / 0.278	58 / 0.403	40 / 0.124	150 / 0.423
Doctors	3 / 0.056	17 / 0.118	6 / 0.019	21 / 0.059
	M= 54	M= 144	M= 322	M= 355

Table C-19
Meetings Attended by IEEE Group Personnel

Meetings Attended (Past Year):	IECI	Circuit Theory	Power	Computer
None	9 / 0.170	34 / 0.236	34 / 0.106	95 / 0.268
One	9 / 0.170	36 / 0.250	36 / 0.112	86 / 0.242
Two	8 / 0.151	28 / 0.194	59 / 0.183	82 / 0.231
Three	10 / 0.189	21 / 0.146	51 / 0.158	31 / 0.087
Four or More	17 / 0.321	25 / 0.174	142 / 0.441	61 / 0.172
	M= 53	M= 144	M= 322	M= 355

Table C-20

Papers Published by IEEE Group Personnel

Papers Published (5 Years):	IECI	Circuit Theory	Power	Computer
None	28 0.519	78 0.542	195 0.606	207 0.583
One	9 0.167	20 0.139	52 0.161	53 0.149
2-3	11 0.204	24 0.167	48 0.149	64 0.180
4-6	5 0.093	13 0.090	23 0.071	22 0.062
Seven or More	1 0.019	9 0.062	4 0.012	9 0.025
	M= 54	M= 144	M= 322	M= 355

Table C-21

Patents Filed by IEEE Group Personnel

Patents Filed (5 Years):	IECI	Circuit Theory	Power	Computer
None	26 0.481	88 0.615	290 0.903	222 0.625
One	6 0.111	20 0.140	12 0.037	54 0.152
2-3	16 0.296	22 0.154	9 0.028	53 0.149
4-6	6 0.111	10 0.070	7 0.022	17 0.048
7-15	0 0.	3 0.021	3 0.009	8 0.023
Sixteen or More	0 0.	0 0.	0 0.	1 0.003
	M= 54	M= 143	M= 321	M= 355

TABLE C-22

Periodicals Read by IEEE Group Personnel

Periodicals Read Regularly:	IECI	Circuit Theory	Power	Computer
None	0 0.	2 0.015	2 0.006	5 0.014
One	0 0.	2 0.C15	6 0.019	13 0.038
Two	4 0.077	8 0.058	23 0.073	27 0.078
3-5	18 0.346	53 0.387	151 0.481	145 0.419
Six or More	30 0.577	72 0.526	132 0.420	156 0.451
	M= 52	M= 137	M= 314	M= 346

Appendix D

The Data

With the notable exception of the fraction of members in a scientific field, the score distributions on all variables, including the dependent, are approximately normal. As a simple test of normality, the actual frequencies of scores on each variable, grouped in eight intervals of one-half standard deviation each (i.e., so that there are four on either side of the mean, with the first including scores more than 1.5 standard deviations above the mean and the eighth, 1.5 below it), were compared with the frequencies expected given normal distribution of the scores. Table D-1 gives those frequencies for the dependent

TABLE D-1

Normality Tests

Group score differs from mean by:	Expected Frequency N = 86	Actual Frequency	
		Local Sources	Scientific Profession
+1.5 standard deviations or more	5.7	7	
+1.01 to 1.5 standard deviations	7.9	11	26
+.51 to 1.0 standard deviations	12.9	12	3
+.0 to .50 standard deviations	16.5	9	2
−.01 to .50 standard deviations	16.5	18	11
−.51 to 1.0 standard deviations	12.9	13	44
−1.01 to 1.5 standard deviations	7.9	12	
−1.5 standard deviations or more	5.7	4	

TABLE D-2

Correlation Matrix for Group Scores

		X(1)	X(2)	X(3)	X(4)	X(5)	X(6)	X(7)	X(8)	X(9)	X(10)	X(11)	X(12)	X(13)	Dependent	
X(1)	Discipline Experience	1.00	0.70	0.22	-0.02	-0.03	0.35	0.21	0.31	0.22	-0.24	0.13	0.24	0.30	-0.11	(1)
X(2)	Application Experience		1.00	0.24	-0.04	-0.04	0.22	0.18	0.27	0.14	-0.21	0.04	0.08	0.18	-0.08	(2)
X(3)	Scientific Profession			1.00	-0.01	0.12	0.74	0.60	0.65	0.43	-0.25	0.18	0.58	0.35	-0.51	(3)
X(4)	Low Seniority				1.00	-0.58	0.01	-0.16	-0.08	-0.18	0.04	-0.02	-0.11	-0.12	-0.06	(4)
X(5)	High Seniority					1.00	0.04	0.21	0.22	0.03	0.01	-0.06	0.22	0.17	0.06	(5)
X(6)	Higher Education						1.00	0.55	0.66	0.39	-0.10	0.30	0.64	0.34	-0.50	(6)
X(7)	Meetings Attended							1.00	0.55	0.37	-0.11	0.32	0.55	0.24	-0.46	(7)
X(8)	Papers Published								1.00	0.32	-0.18	0.34	0.56	0.26	-0.40	(8)
X(9)	Patents Issued									1.00	-0.22	0.04	0.46	0.16	-0.21	(9)
X(10)	Recent Instance										1.00	0.02	-0.16	-0.10	0.14	(10)
X(11)	Scientific Task											1.00	0.38	0.21	-0.38	(11)
X(12)	Many Periodicals												1.00	0.31	-0.55	(12)
X(13)	Competence Mode													1.00	-0.38	(13)

variable, which is approximately normal, and for "scientific professional field," which is not. Among the 1,204 scores computed (14 variables for each of 86 groups) only four cases fell more than three standard deviations from the mean for the variable in question.

Factor Analysis

As Table 3-3 shows, multiple regression on scores from all 86 groups produced an estimating equation which excludes the scores for higher education, meeting attendance, and papers published, although any one of these scores could explain from 15% to 25% of the variance in the dependent variable. At the same time, the equation includes "high seniority," which on a marginal basis accounts only for a fraction of 1% of the original variance. The complete correlation matrix, in Table D-2, shows that there is a high degree of intercorrelation among the five variables associated with high degree of commitment to the development of competence. Those five include the two given greatest weight in the estimating equation and the three whose absence was just noted. Because of these intercorrelations, we carried out a factor analysis of the grouped data to see if a smaller number of variables could provide greater predictive utility.

By factor analysis the 12 independent variables (excluding the control for competence-oriented acquisition) were reduced to five factors, using the method of principal-components analysis and the oblimin criterion for rotation.[1] The standardized factor weights are given in Table D-3. Table D-4 presents the seven-variable correlation matrix (the dependent variable, five factors, and the control). The first factor we call "professional commitment," since it is associated with the measures of educational level, meeting attendance, publications, periodical readership, and inventive achievement. The other factors are tied either to a single score or to a pair of closely related scores; hence descriptive terms were easily chosen for them.

Multiple regression on three factor scores plus the control for competence mode will explain 42% of the variance in the incidence of use of company sources among the 86 groups. The estimating equation, beta weights, and standard errors for the betas are given in Table

---

[1] This was carried out, as were the other multivariate tests described in Chapter 3, by the Multivariate Statistical Analyzer (MSA) Program at the Harvard University Computing Center. See K. Jones, "The Multivariate Statistical Analyzer," 1964; W. W. Cooley and P. R. Lohnes, *Multivariate Procedures for the Behavioral Sciences*, 1962; and H. H. Harman, *Modern Factor Analysis*, 1960, Chapters 9 and 15.

Table D-3

Factor Analysis of Independent Variable Scores for 86 Groups*

	Mean Score	Standard Deviation	Factor Weights on Normalized Row Vectors				
			1 Professional Activity	2 Seniority	3 Experience	4 Interval	5 Task
Variable							
X₁ Discipline Experience	.53	.21	.17	−.02	94	−.10	.04
X₂ Application Experience	.71	.17	.07	.01	.98	−.05	−.07
X₃ Scientific Profession	.37	.43	.97	−.07	.05	−.04	−.13
X₄ Low Seniority	.24	.17	.02	−.99	−.04	.11	−.01
X₅ High Seniority	.34	.23	.14	.97	−.06	.14	−.09
X₆ Higher Education	.34	.26	.96	−.14	.14	.14	.03
X₇ Meetings Attended	.68	.20	.92	.20	.04	.05	.17
X₈ Papers Published	.37	.25	.90	.10	.20	.05	.18
X₉ Patents Issued	.32	.26	.76	.07	−.01	−.41	−.36
X₁₀ Recent Instance	.81	.14	−.06	.02	−.15	.97	−.06
X₁₁ Scientific Task	.20	.20	.25	−.06	−.01	−.05	.94
X₁₂ Many Periodicals Read	.29	.23	.92	.12	−.08	−.10	.25

* The five factors defined here account for 77% of the variance in the original data.

Table D-4

Factor Scores

Means, Standard Deviations, and Correlation Matrix:

| | | Mean | Standard Deviation | Correlation Matrix (Pearson r) | | | | | | |
				$F1$	$F2$	$F3$	$F4$	$F5$	X_{13}	Dep.
Factor 1	Professional Commitment	0.0	.93	1.0	.09	.17	−.13	.11	.36	−.57
Factor 2	(Seniority)	0.0	.89		1.0	−.01	−.03	.02	.14	.08
Factor 3	(Experience)	0.0	.91			1.0	−.05	.05	.22	−.04
Factor 4	(Recent)	0.0	.92				1.0	.18	−.05	.09
Factor 5	(Task)	0.0	.95					1.0	.14	−.26
X_{13}	Competence	.13	.12						1.0	−.38
Dependent Variable	Company Source	.51	.25							1.0

TABLE D-5

Multiple Regression Analysis: Incidence of Company Sources, on Five Factors, Controlled for Competence Mode
(86 Groups)

Regression equation for estimating fraction of sources within the same company:

$\hat{Y} = -.132$ F1 (Pro. Activity) $-.050$ F5 (Task) $+.046$ F2 (Seniority) $-.429$ X_{13} (Competence Mode) $+.570$

Multiple r $= .649$ Multiple r^2 $= .42$

	Beta Weights	Standard Errors of Beta Weights
F1	−.493	.091
F2	.163	.086
F5	−.187	.086
X_{13}	−.201	.092

D-5. Comparison with Table 3-3 shows that using either method, the best predictors are the measures of professional commitment, with the others carrying lesser and approximately equal weights.

Sorting the Groups

We can look more closely at the composition of groups and the nature of their work setting. As one would expect, groups tend to be composed predominantly of men in a single field. Although the mean membership fraction for scientists is .36, 20 groups contained only scientists, and 37 contained none at all. It seemed reasonable then to separate the groups into two classes: "scientific" and "engineering." [2] A tabulation of groups by location and type is given in Table D-6.

TABLE D-6

Groups by Location and Type

Location		Number of Groups by Type		
Business	Establishment	Scientific	Engineering	Total
MEDCO		19	1	20
BASIC *		1	3	4
POLY	A		2	2
	B	4		4
	C	5	1	6
MECHO		1	14	15
CONTROL	A		4	4
	B		11	11
EDP			20	20
		30	56	86

* All four groups were in a second, smaller research establishment, which shared the same facilities with the main BASIC laboratory, for which organizational affiliations were unavailable.

By developing separate estimating equations for the "scientific" and "engineering" groups (which also tend to be located, respectively, in central laboratories and operating divisions), we gain some slight im-

---

[2] Heterogeneous groups with more than 40% scientists were called "scientific"; the rest were classed as "engineering." Of the 19 "engineering" groups with some scientific membership, 15 had less than 25% in that category. Nine of these 19 heterogeneous groups were in the EDP organizational of DELTA Corporation.

provement, being able to explain 50% of the original variance. The sums of the squared deviations from the mean for the dependent variable are analyzed in Table D-7.

TABLE D-7

Variance in Group Score for Dependent Variable
(*Company Sources*)

	Original Sum of Squares	Explained by Regression	Residual
		(Table 3-3)	
All 86 Groups	5.40	2.44	2.96
30 "Scientific"		(Table D-8)	
Groups	1.13	.42	.71
56 "Engineering"		(Table D-9)	
Groups	3.01	1.00	2.01

Table D-7 shows that 22% of the variance is accounted for by the simple classification of group composition (slightly more is accounted for by retaining the original measure of composition, since the correlation coefficient is .51). Conditional analysis of the two smaller groups produced the results summarized in Tables D-8 and D-9. Controlling for competence mode, four variables account for 34% of the residual variance in the 56 engineering groups. The best predictor again is periodical readership, the others being meeting attendance — which is another measure of professional commitment —

TABLE D-8

Regression Analysis: Incidence of Company Sources, 30 Scientific Groups

Estimating Equation:

$\hat{Y} = .206\ X_5$ (Seniority) $+ 226\ X_9$ (Patents) $- .309\ X_{11}$ (Scientific Task) $- .639\ X_{13}$ (Competence Mode) $+ .358$

Multiple $r = .612$ Multiple $r^2 = .37$

	Beta Weights	Standard Errors of Beta Weights
X_5	.222	.167
X_9	.338	.173
X_{11}	−.355	.163
X_{13}	−.388	.170

TABLE D-9

Regression Analysis: Incidence of Company Sources,
56 Engineering Groups

Estimating Equation:

$\hat{Y} = .311\ X_5$ (Seniority) $- .224\ X_7$ (Meetings) $- .496\ X_{12}$ (Periodicals) $- .635\ X_{13}$ (Competence Mode) $+ .800$

Multiple $r = .577$ Multiple $r^2 = .33$

	Beta Weights	Standard Errors of Beta Weights
X_5	.324	.120
X_7	−.178	.129
X_{12}	−.394	.131
X_{13}	−.290	.115

high seniority, and scientific task. Results for the 30 scientific groups are reported for the sake of completeness, but their validity is doubtful. The small sample size increases the risk of misinterpreting random associations, while the fact that nearly two-thirds of the cases come from only one laboratory — MEDCO — limits the interpretation of any meaningful results. There is one surprise in the results, since this

TABLE D-10

Summary of Mean Scores and Correlations
with the Dependent Variable

	86 Groups		30 Scientific Groups		56 Engineering Groups	
	mean	r	mean	r	mean	r
X1 Discipline Experience	.53	−.11	.59	.07	.50	−.05
X2 Application Experience	.71	−.08	.76	.04	.69	.00
X3 Scientific Field	.37	−.51	.93	.13	.07	−.38
X4 Low Seniority	.24	−.06	.24	−.03	.23	−.07
X5 High Seniority	.34	.06	.38	.17	.31	.14
X6 Higher Education	.34	−.50	.59	.06	.20	−.38
X7 Meetings	.68	−.46	.84	−.12	.60	−.30
X8 Papers Published	.37	−.40	.57	−.06	.26	−.21
X9 Patents Filed	.32	−.21	.46	.23	.24	−.17
X11 Scientific Task	.20	−.38	.24	−.44	.17	−.31
X12 Periodicals Read	.29	−.55	.45	−.38	.19	−.40
X13 Competence Mode	.13	−.38	.19	−.26	.10	−.26
Dependent, Company Source	.51	1.0	.35	1.0	.60	1.0

is the only case in any of our analyses in which patent statistics proved to be related to the means of information transfer. In view of the considerations stated, we place little weight on this outcome, although it may signify something about the MEDCO organization or perhaps industrial work in the life sciences in general.

Table D-10 summarizes mean scores and correlations with the dependent variable for all 86 groups, and the scientific and engineering groups separately.

Bibliography

Allen, Thomas J., "Performance of Information Channels in the Transfer of Technology," *Industrial Management Review*, Vol. 8 (1966), pp. 87–98.

———, "Sources of Ideas and Their Effectiveness in Parallel R&D Projects," Working Paper 130–65, Sloan School of Management, Massachusetts Institute of Technology, 1965.

Auerbach Corporation, "DOD User Needs Study, Phase I," Final Technical Report, 1151-TR-3, 2 Vol., Philadelphia, Pa., May 1965.

Barnes, L. B., *Organizational Systems and Engineering Groups.* Boston: Division of Research, Harvard Business School, 1960.

Bauer, R. A., "Problem Solving in Organizations: A Functional Point of View," in *Business Policy Cases with Behavioral Sciences Implications*, M. M. Hargrove, I. H. Harrison, and E. S. Swearingen, editors. Homewood, Illinois: Richard D. Irwin, 1963.

———, "The Obstinate Audience," *American Psychologist*, Vol. 19 (1964), pp. 319–328.

Bruner, J. S., J. J. Goodnow, and G. A. Austin, *Study of Thinking.* New York: Science Editions, 1962.

Committee on Scientific and Technical Information (COSATI), *Recommendations for National Document Handling Systems in Science and Technology*, Washington, D.C., 1965.

Cooley, W. W., and P. R. Lohnes, *Multivariate Procedures for the Behavioral Sciences.* Now York: John Wiley, 1962.

Crawford, J. H., G. Abdian, W. Fazar, S. Passman, R. B. Stegmaier, Jr., and J. Stern, *Scientific and Technical Communications in Government*, Task Force Report to the President's Special Assistant for Science and Technology, April 1962 (AD-299-545).

Cuadra, C., editor, *Annual Review of Information Science and Technology*, Vol. I. New York: John Wiley, 1966.

Federal Council for Science and Technology, *Proceedings*: Second Symposium on Technical Information and the Federal Laboratory, U.S. Government Printing Office, 1964.

Festinger, L., and D. Katz, *Research Methods in the Behavioral Sciences*. New York: John Wiley, 1953.

Ghiselli, E. E., *Theory of Psychological Measurement*. New York: McGraw-Hill, 1964.

Glaser, B. G., "The Local-Cosmopolitan Scientist," *American Journal of Sociology*, Vol. 69 (1963), pp. 249–259.

Goldberg, L. C., F. Baker, and A. H. Rubenstein, "Local-Cosmopolitan: Unidimensional or Multidimensional," *American Journal of Sociology*, Vol. 70 (1965), pp. 704–710.

Gouldner, A. W., "Cosmopolitans and Locals: Toward an Analysis of Latent Social Roles," *Administrative Science Quarterly*, Vol. 2 (1951), pp. 281–306, 444–480.

Halbert, M. H., and R. L. Ackoff, "An Operations Research Study of the Dissemination of Scientific Information," *ICSI* * (1959), pp. 97–130.

Hanson, C. W., "Research on Users' Needs: Where Is It Getting Us?" *Aslib Proceedings*, Vol. 16 (1964), pp. 64–78.

Harman, H. H., *Modern Factor Analysis*. Chicago: University of Chicago Press, 1960.

Herner, S., "Information Gathering Habits of Workers in Pure and Applied Science," *Industrial and Engineering Chemistry*, January 1954, pp. 228–236.

Jones, K., "The Multivariate Statistical Analyzer," Harvard University, 1964. (Mimeo.)

Martin, M. W., "The Measurement of Value of Scientific Information," in *Operations Research In Research and Development*, B. V. Dean, editor. New York: John Wiley, 1963, pp. 97–123.

Marvick, D., *Career Perspectives in a Bureaucratic Setting*, Michigan Governmental Studies No. 27. Ann Arbor: University of Michigan Press, 1954.

McLaughlin, C. P., R. S. Rosenbloom, and F. W. Wolek, "Technology Transfer and the Flow of Technical Information in a Large Industrial Corporation," Harvard University, 1965. (Mimeo.)

* ICSI refers to Volume I of the *Proceedings* of the International Conference on Scientific Information. Washington, D.C.: National Academy of Science, 1959.

Menzel, Herbert, "Can Science Information Needs Be Ascertained Empirically?" *Communication: Concepts and Perspectives (Proceedings of the Second International Symposium on Communication Theory and Research)*, Lee Thayer, editor. Washington: Spartan Books, 1966 (b).

————, "The Flow of Information Among Scientists," Columbia University, 1958. (Mimeo.)

————, "Information Needs and Uses in Science and Technology," in Cuadra, *op. cit.*, pp. 41–69 (1966) (d).

————, "The Information Needs of Current Scientific Research," *Library Quarterly*, Vol. 34 (1964), pp. 4–19.

————, "Review of Studies in the Flow of Information Among Scientists," Columbia University, Bureau of Applied Social Research, 1960. (Mimeo.)

————, "Scientific Communication: Five Sociological Themes," *American Psychologist*, Vol. 21 (1966), pp. 999–1004 (a).

Morton, J., "From Research to Technology," *International Science and Technology*, May 1964, pp. 82–92.

National Science Foundation, *Basic Research, Applied Research, and Development in Industry*, 1965, NSF 67–12, 1967.

North American Aviation, Inc., "DOD User Needs Study, Phase II," Final Report, C6-2442/030, 2 Vols., Anaheim, California, November 1966.

Organization for Economic Cooperation and Development, *Technical Information and the Smaller Firm*. Paris: O.E.C.D., 1960.

Paisley, W. J., "The Flow of (Behavioral) Science Information — A Review of the Research Literature," Stanford University, Institute for Communications Research, 1965. (Mimeo.)

Pelz, D. C., "Some Social Factors Related to Performance in a Research Organization," *Administrative Science Quarterly*, Vol. 1 (1956), pp. 310–325.

President's Science Advisory Committee, *Science, Government, and Information*. Washington, D.C.: Government Printing Office, 1963.

Price, D. J., de S., "Is Technology Historically Independent of Science?" *Technology and Culture*, Vol. 6 (1965), pp. 553–568.

Scott, C., "The Use of Technical Literature by Industrial Technologists," *IEEE Transactions on Engineering Management*, Vol. EM-9 (1962), pp. 76–86.

————, with L. T. Wilkins, *The Use of Technical Literature by Industrial Technologists*. London: The Social Survey, Revised Edition, 1960.

Shepherd, C. R., and P. Brown, "The Impact of Altered Objectives: Factionalism and Organizational Change in a Research Laboratory," *Social Problems*, Vol. 34 (1956), pp. 235–243.

Taube, M., "An Evaluation of Use Studies of Scientific Information," Documentation, Inc., December 1958 (AD 206987).

79